THE HOLE IN THE UNIVERSE

Also by K. C. Cole

First You Build a Cloud

The Universe and the Teacup

THE HOLE IN THE UNIVERSE

How Scientists Peered over the Edge of Emptiness and Found Everything

K. C. COLE

HARCOURT, INC.

New York San Diego London

www.harcourt.com

Library of Congress Cataloging-in-Publication Data
Cole, K. C.
The hole in the universe: how scientists peered over the edge of emptiness and found everything/by K. C. Cole.—1st ed.
p. cm.
Includes bibliographical references and index.
ISBN 0-15-100398-X
1. Physics—Philosophy. 2. Nothing (Philosophy). I. Title.
QC6 .C62 2000
530'.01—dc21 00-044947

Illustrations by David Barker

Designed by Lois Stanfield, LightSource Images
Text set in Garamond 3
Printed in the United States of America
First edition
A C E G I J H F D B

For Liz,
Millennium Girl,
Congratulations
and Love,
Mom

Contents

Appreciations, Attributions, and Apologies

Nothing is far and away the most difficult subject I have attempted to pin down on the pages of a book. On the one hand, it's so boundless that it refuses to stay put, oozing into everything. On the other hand, it can be so soft and insubstantial that it eludes a hard look. Grasping "nothing" requires resisting the temptation to follow it wherever it leads, getting lost in the semantic thicket of nothing puns, or simply bouncing the idea around on one's knee, stringing together curious facts and ancient history—taking it for a pleasurable, if rather pointless, trip.

I have tried to avoid these temptations, and to the extent I have failed, I hope the reader will forgive me. I hope I have managed, after all, to say *something* about nothing. At the least, I hope to convince the reader that nothing matters. Most of modern physics and mathematics would be unthinkable without it. The human mind we use to perceive and explore those worlds relies on notions of "nothing" in very tangible ways—creating something out of nothing as handily as it does the reverse.

In one way or another, I have been writing about "nothing" for nearly twenty years—ever since I first heard physicists casually tossing about the term "vacuum" at meetings in a way that made it clear they were talking about a real, palpable, and critically important phenomenon. Since that time, I've been pursuing the subject as the physical science writer for the *Los Angeles Times.* I thank the *Times* for that opportunity—especially my editor, Joel Greenberg—and also for leave in which to write the book.

The material for this book was gathered primarily from three broad categories of sources: books and papers, meetings, and interviews. The books and papers are listed in the bibliography and often cited in the text as well. Meetings are also sometimes cited in the text. In cases where no source is cited, readers can assume that direct quotations are based on personal conversations of one kind or another.

I have omitted notes because statements that read as "fact" are at the very least the consensus of a wide range of experts. In places where a discovery or a turn of phrase or an idea seems the province of one person more than others, I've stated that person's name. Where consensus is difficult—for example, in cosmology or string theory—I have tried to indicate whether the ideas discussed are widely accepted, highly tentative, or merely idiosyncratic. In cases where a discovery or idea is attributed to a specific scientist, chances are it was not a solo effort; the reader would probably be safe to automatically add "and colleagues" after most attributions.

The book covers a great deal of ground in relatively few words and focuses on tracking ideas rather than those who deserve credit for them. In fast-forwarding through important historical periods and intellectual developments, it leaves out a great deal. To make up for these omissions, I have tried to include in the bibliography (and sometimes in the text) what I feel is the best accessible source on the subject at hand. Luckily, there are many excellent accounts: for example, Brian Greene on string theory, Alan Guth on cosmology, Kip Thorne on general relativity.

Most important of all were the many conversations over the years with scientists courageous enough to venture into the void. Some of those are listed in the supporting cast—along with other people I cite frequently throughout the book. I would like to thank all of them—as well as those I have forgotten to mention—for their insights and time. I also thank the Mathematical Sciences Research Institute in Berkeley and the Exploratorium in San Francisco for summer fellowships that made much of the research possible. Napoleon Williams gets a special thanks for his continual flow of nothing tidbits and thoughts and for his general enthusiasm for this project.

Some scientists went far beyond mere conversation and also read sections of this book as it was coming to be, offering comments, criticism, and suggestions. Their efforts have vastly improved what follows. In particular, I wish to thank Brian Greene, Rocky Kolb, Haim Harari, Lynn Cominsky, Alan Guth and Richard Brown.

A very special thanks to physicist Thomas Humphrey—artist, musician, and master physics teacher—for plowing through the entire manuscript and for many entertaining and enlightening conversations about everything and nothing over the past twenty-five years.

I owe an enormous debt of gratitude to Tom Seigfried, science editor of the *Dallas Morning News* and author of the charming and important new book, *The Bit and the Pendulum: From Quantum Computing to M Theory—The New Physics of Information.* Tom scoured every line of this manuscript and found numerous mistakes and inconsistencies. I don't know anyone else who manages to combine such a fierce commitment to accuracy with such a fine clarity of expression. Speaking of sticklers, many thanks to my patient and thorough and good-humored copy editor, Rachel Myers.

Susan Chace and Patty O'Toole—superb writers and esteemed colleagues in all things—offered early and critical advice, direction and support.

As always, I relied enormously on my friends and family, my agent Ginger Barber, my courageous colleagues at JAWS (Journalism and Women Symposium) and PEN Center USA West.

I wish to acknowledge the inspiration of my Spring 2000 Concepts of Nothing Honors Collegium students at UCLA, and especially Jordan Richmond, from whom I stole the idea of a "hole in the heart" that opens the book. I thank Stan Wojcicki for pointing me to the story from Richard Feynman that closes the book. Finally, a big thanks to David Barker for the inspired illustrations.

The inspiration for the book came from Jane Isay, my editor, publisher, and biggest fan, who asked me if I'd like to write a book about a number. Zero came immediately to mind. Nothing followed. Thanks for nothing, Jane.

THE HOLE IN THE UNIVERSE

Chapter 1

WHY NOT? A PRELUDE

Nothing is too wonderful to be true.
—MICHAEL FARADAY

THERE IS A HOLE in the universe.

It is not like a hole in a wall where a mouse slips through, solid and crisp and leading from somewhere to someplace. It is rather like a hole in the heart, an amorphous and edgeless void. It is a heartfelt absence, a blank space where something is missing, a large and obvious blind spot in our understanding of the universe.

That missing something, strange to say, is a grasp of nothing itself. Understanding nothing matters, because nothing is the all-important background upon which everything else happens. Without it, the universe is theater without a stage. Without getting to know it, we can't understand the blank page on which the story of everything is written. We can't trust our own perceptions because everything we see passes through it like a clear but distorting lens, like light from the sky skidding over hot pavement to create a shimmering mirage.

For centuries, scientists, mathematicians, and philosophers have tried to track nothing down, give it a name, put a box around it.

They've dressed it up with all kinds of decorative effects, like those daisy decals old folks in Florida stick on their sliding glass doors, the better to see the invisible, to avoid bumping their heads. But nothing continues to fool them. Wherever they go, they bump up against it.

Today, nothing is back with a vengeance, at the forefront of everything. It is the font of all creation—a hyperactive busybody that expands, explodes, spawns, wiggles, stretches, curls, twirls, pops, burrows, shakes things up, and generally interferes with everything. And yet, it remains as elusive as ever, the chameleon at the center of the cosmos.

How can such a powerful presence remain so effectively masked? And how can nothing have such profound effects?

To imagine how it might happen, consider a very ordinary kind of "nothing"—like a blank piece of paper. When you write or draw on this featureless background, you're free to create anything you like. From scratch. Ex nihilo. A perfectly featureless background like a blank piece of paper couldn't possibly affect what we draw on it. Or so we like to think.

But what if...

The paper is bumpy so that any mark you draw on it skips and sputters from place to place, and you find that it's impossible to draw a perfectly smooth line.

Or the paper is slippery, so that your pen slides and the ink oozes off the edge.

Or the paper is curled into a cylinder, so that even a straight line circles around and meets itself from the rear.

Or the paper is black—so anything you draw on it disappears.

Or the paper is three-dimensional, like a cardboard box: suddenly you have many more possibilities for what you can create.

Or the paper is one-dimensional, like a line: your possibilities are constricted.

Or the paper has zero dimensions, or ten, and they are knotted and twisted in bizarre ways.

Or the paper wiggles and waves as you try to write on it. It won't stand still.

Or the paper has a barely perceivable background, an intricate set of images that you couldn't see until you developed the right technology.

Or the paper grows, stretches, shrinks, changes shape before your eyes.

Or the paper itself starts to draw lines and figures of its own accord.

While these scenarios may seem bizarre, they are not so different from those that have faced mathematicians and physicists in their search "for nothing at all." The properties of nothing don't always show their colors directly. That's why physicists have to do clever experiments (and thought experiments) to ferret them out.

Think of a large rock, perched on the edge of a high cliff, doing "nothing." On further inspection, you find that the nothing it's doing includes hurtling through space at enormous speeds, carried along by the spin of the earth, the orbit of the earth around the sun, the motion of the sun around the pinwheel of the galaxy.

In addition, the rock's "nothing" contains enormous potential. It's poised to act. If it falls off the cliff and crashes to the bottom, that potential will turn into enough energy to crush the skull of someone passing underneath at the wrong time.

Potential, it turns out, is one of the most impressive properties of nothing.

There are hints that nothing eludes us precisely because it is too featureless, like perfectly transparent glass. Perhaps "nothing" is perfect—too perfect to perceive. Or in Faraday's words, "too wonderful to be true." To make something from nothing, we must crack the glass, destroy the symmetry. When nothing shatters, everything can be born.

SWEET NOTHING

Anybody who knows all about nothing knows everything.

—physicist LEONARD SUSSKIND, Stanford University

From our earliest days, we've grown accustomed to thinking of nothing as a kind of bland absence—a convenient pause between numbers or atoms or thoughts, a passive-aggressive empty space that resembles nothing so much as a blank stare.

Nobody home. Nothing doing. Nothing on my mind.

Nothing could be further from the truth.

In the past few hundred years, the struggle to get a handle on nothing has changed the course of mathematics, physics, and even the study of the human mind. And while that's a fact well known to science, precious few laypeople have been let in on the secret: While no one was looking, nothing became a central player that creates number systems out of whole cloth; bubbles up matter and universes; materializes sights, sounds, perceptions. As physicists and mathematicians plunged deeper into the void, they emerged with an abundance of riches that seems to have no end in sight. Indeed, the evolution of nothing into a full-fledged player in the universe stands as one of the single greatest paradigm shifts in human thought.

As physicist James Trefil has put it, "Nothing just ain't what it used to be."

It's hard to understand how nothing got such a bad name in the first place. Ancient Greek thinkers spread the rumor that nature abhors a vacuum. There is no evidence of this, as modern physicists frequently point out. It is rather *people* who appear to abhor the vacuum. We are taught from childhood to shun the shadowy, dread the dark side, fill the void. People feel compelled to plug gaps in conversation, devise activities to do something (anything!) about dead (which is to say "empty") time. We describe the deranged as "not all there." We consider it shameful to think negative. No one loves a lack.

And yet, nothing may be the single most prolific idea ever to plop into the human brain.

Consider the simple naught—the zero, the goose egg, the zilch—a precocious offspring if ever there was one. Conceived as a simple placeholder to note what was not, it started misbehaving almost as soon as it appeared on the scene—confounding, confusing, creating paradoxes left and right. It grew into an unruly monster that brought instant death by multiplication, wrought absurdity through division, exploded into the cloudy ambiguity of infinity. At the same time, it opened the gates to whole new realms of numbers, including negatives, imaginaries, and infinitesimals so ephemeral they were once dismissed as "ghosts."

To close in on zero is to slide down a slippery slope, but at the bottom is an undreamed-of bounty—a cache of intellectual tools so powerful they lie behind virtually all of physics, philosophy, technology, and higher mathematics. Indeed, every number—and every conceivable number system—can be created from the humble origin of that ultimate egg.* For good and for ill, the invention of zero let the genie—and also the genius—out of the bottle.

In physics, the study of nothing lies at the bottom of every burning question from the cramped quarters of subatomic spaces to the expansive realm of the cosmos at large, and especially at the ragged edges where the largest and smallest meet. All properties of matter, of forces, of space, and of time are intricately woven into the vacuum itself—that is, the ultimate nothing. Particle physicists try to

*See Chapter 3, "Good for Nothing."

melt the vacuum or to catch pieces of it in their detectors, while astronomers try to snare waves of spacetime in laser nets. Theorists would like nothing better than to understand the unexplainable lightness of nothing. When theorists calculate the energy in nothing, they come up with enormous numbers. The energy in nothing should be huge. And yet, gravity doesn't seem to know it's there. Gravity has "weighed" the ever-increasing nothingness in the universe and found it lacking.

The loveliness of mathematics and physics is that it allows us to move the search for nothing out of the realm of pure navel gazing and into an arena where concrete questions can be posed.

In a broader context, absences loom large in our attempts to understand just about anything. For example, what we don't notice tells us a great deal about how our brains perceive the world of things, people, and relationships "out there" in what we like to call the real world. We do not see the thick sustenance-giving web of blood vessels that veils our visual field; we don't feel the clothes on our backs or just about anything that isn't the central focus of our fickle attentions.

In fact, what we don't say, don't hear, don't feel, don't remember, don't ask, don't tell, in itself tells psychologists and neuroscientists more about the human mind and the tangled web of neurons and culture that creates it than perhaps any other category of evidence. We define ourselves by what we're not as much as by what we are. The physicist I. I. Rabi defined himself as an Orthodox Jew because, as he put it, that was the church he was *not* attending.

These ever-present absences also inform the work of historians, writers, philosophers, artists.

By chance, I shared an office at the Exploratorium in San Francisco briefly in the summer of 1999 with New York artist Fred Wilson. Fred has just won a MacArthur fellowship for his installations that reveal what museums *don't* show.

"What they put on view says a lot about the museum," he told me, "but what they don't put on view says more."

He created one of his first major installations at the Maryland Historical Society. Along with silver tea services, the society's exhibits included busts of "great men" on pedestals, with suitably dignified labels: Henry Clay. Andrew Jackson. Napoleon. "Most of them weren't even from Maryland," Fred said.

Conspicuously missing were figures central to Maryland's history: Harriet Tubman, Frederick Douglass, and Benjamin Banneker (a black mathematician who surveyed Washington, D.C., for Thomas Jefferson). So Wilson added a series of pedestals with labels, but no heads.

The absence spoke loud and clear—and far more forcefully than a presence ever could.

In the same way, playwright Samuel Beckett—a man of few words if ever there was one—prolifically employed thoughts left unspoken in his plays. Language, he said, was like a veil that obscured meaning. "[Language] must be torn apart in order to get at the things (or the Nothingness) behind it." His technique, he said, was to take away, "subtracting rather than adding."

NOT

Every craftsman searches for what's
not there to practice his craft.

—thirteenth-century poet RŪMĪ, *Work and Emptiness*

The doors to scientific breakthroughs are often just such holes in the understanding, gaps in the data. Biologists study missing links in the chain of evolution, missing branches in the family tree, missing hair, tails, chromosomes. Chemists have discovered more than one element as a gap in the periodic table. Physicists, in particular, spend a good portion of their time investigating the properties of things that can't be seen, don't matter, don't exist, or are generally AWOL.

Take the notorious case of the missing matter. Astronomers can clearly see that galaxies spin around at speeds fast enough to send the stars flying into space like water droplets off a salad spinner. Yet the galaxies somehow hold the stars together in formation—so many sparkling stellar sparrows. The only force that could glue the galaxies together is gravity. Gravity, however, comes from mass, and astronomers see only a small portion of the matter necessary to do the trick. The rest can't be found.

Or consider the missing neutrinos. Astrophysicists have a pretty good idea what makes the Sun shine—nuclear fusion. But these well-understood fusion reactions should shower Earth with untold numbers of practically nonexistent particles called *neutrinos*—"little neutral ones." Trouble is, only a third to a half of the predicted number of them actually seem to arrive on Earth. The rest get lost on their journey from the Sun to the scientists' detectors. Physicists think neutrinos may be cosmic shape-shifters, changing from a detectable form to another harder-to-detect species en route.

Other missing entities abound—magnetic monopoles, for example. These never-seen particles would be magnetism's counterpart to the electron—a single electric charge. The negatively charged electron can stand on its own, as can the positively charged proton. But magnets, for some reason, always come with north and south poles permanently attached. Why aren't there single north or single south poles?

The larger question is why scientists waste their time studying what can't be found when so much is right in front of their noses.

One answer is that finding the missing pieces helps to prove—or disprove—the theories that suggest these entities should exist in the first place. In the past, physicists have discovered antimatter, quarks, and even neutrinos through just such searches.

But even searches that come up empty-handed are uncannily useful. In Einstein's day, physicists were busily speculating about the properties of the luminiferous ether—the pervasive medium that everyone assumed carried waves of light like air carries waves of wind. Clever experiments revealed that the ether did not exist. Einstein's revolutionary insights on the nature of light, space, and time later explained why nature didn't require an ether at all.

And so it goes. What's *not* is as significant as what *is.* Mathematicians even study knots by exploring the spaces the knots *don't* occupy. These empty spaces are known in the trade as "not knots."

To physicists, in fact, there's probably nothing more important than things that don't matter, don't happen. For example, the laws of nature seem constructed on a small set of properties—patterns, if you will—that never change. The total amount of electric charge is a constant, as is the total amount of energy or momentum. These "conserved quantities" are the constants in an ever-changing universe, the things that can't disappear no matter what. Every time physicists discover another thing that never changes, they create a new "conservation law" to account for it.

Indeed, the main goal of physics is to discover which properties of the universe survive every kind of alteration, stay the same no matter what you do to them. "As a physicist," said San Francisco's Thomas Humphrey, "you want to discover what doesn't matter at all."

NOT KNOTS

In arithmetic as in politics, the importance of one is
determined by the number of zeros behind him.
—ANONYMOUS SAYING

Part of the power and appeal of nothing springs from this obvious duality. Nothing is *some* thing and *no* thing at the same time. At times, a simple lack. At others, as rich with possibilities as a pregnant pause.

Even in elementary counting, the number zero wears dramatically different hats in different contexts. A person with zero bananas has nothing—banana-wise. Yet the difference between 10 and 10,000,000 (nothing but a string of identical zeros) is huge. In the number 0.01, the first zero means naught; the second turns a tenth into a hundredth.

Zero is both absolute and arbitrary. We count down to zero degrees, descend to ground zero, count backward to zero seconds before blastoff . . . 5, 4, 3, 2, 1 . . . *boom!* But the countdown to the year 2000 has little absolute significance. It's just an arbitrary number some people made up.

The fact that "nothing" can be both something very real and nothing much invites all sorts of double entendres, providing endless fodder for artists, playwrights, novelists, and anyone who enjoys wordplay.

"Nobody was to blame," master math expositor Martin Gardner likes to say. "He usually is."

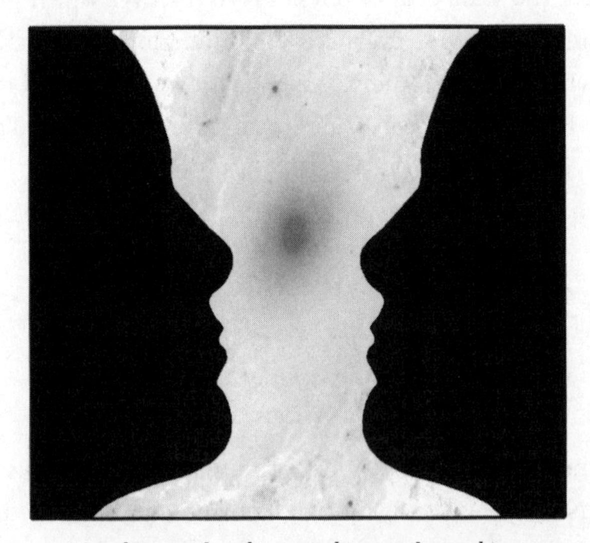

Something and nothing; nothing and something.

Lewis Carroll played similar games. In the Alice adventures, Gardner points out, the White King wonders why Nobody did not arrive ahead of the March Hare, "because nobody goes faster than the hare."

Homer used this same trick in the *Odyssey.* As the one-eyed Cyclops gobbles down Ulysses' crew, our brave hero has the presence of mind to trick the giant into thinking that his name is "Nobody." Later—after Ulysses gets the giant drunk and plunges a burning stake into his single eye—the punctured people-eater can't rally his fellow giants to his aid: When they ask who's tormenting him, the Cyclops answers, in anguish, "Nobody!"

The list goes on. The main player in Shakespeare's *King Lear* is nothing (perhaps he should have called *this* play *Much Ado about Nothing*). It's Lear who utters the famous line: "Nothing can be made out of nothing."

My personal favorite duality that springs from nothing is that ever-two-faced object called the hole—an absence you can get into deeply. Is a hole a presence or an absence? By definition, it's an absence; but without holes, bread wouldn't rise, soda wouldn't sparkle, hemoglobin couldn't carry oxygen through the blood, bees couldn't make honeycombs, doughnuts wouldn't exist. Not to mention Swiss cheese.

A black hole is stranger still. These terminal puckers in spacetime are created when gravity gets carried away by an overabundance of matter, crushing everything in the vicinity into a single point. This so-called singularity thus contains everything that went into it. At the same time, it has no dimension. It's nothing and everything at the same time.

Many ideas in physics, in fact, hinge on the inherent duality of holes. Elementary particles called positrons were originally considered holes in a sea of negative charge. These holes are now better understood as particles of antimatter. When the hole gets "filled" with its proper particle, both particle and hole are destroyed—which is exactly what happens when matter and antimatter meet.

Closer to home, "holes" (missing electrons) drifting through silicon crystals power most of our electronic appliances.

And recently, physicists working on string theory—which views everything in the universe as harmonic vibrations of strings in eleven-dimensional space—have dreamed up scenarios in which holes might hold the explanation for one of the most puzzling aspects of elementary particles.* For reasons unknown, the fundamental building blocks of nature appear to be grouped into "families" with similar properties. Now, it appears that members of the families may be related by the number of "holes" in the knotted landscapes of the extra dimensions.

A HANDLE ON NOTHING

It's a struggle to understand how at one instant there's something and at the next instant, there's nothing.
—cosmologist ROCKY KOLB, University of Chicago

At this point, the reader is probably asking: Well, what is "nothing" anyway? Can it even be defined, or is this all so much semantic foolery? Nothing has baffled people ever since the days when early hominids scratched their hairy heads in contemplation of the cold black sky. Present-day physicists aren't much different. It stands to reason: It is almost impossible to think about nothing without giving it all sorts of properties. And then, of course, it isn't nothing anymore.

Consider a fish. To a fish, nothing might be perfectly still water—a medium so pervasive and featureless that it's all but insensible. Food floats within the water, currents run through it, its

---

*See Chapter 6, "Nothing Gets Strung Out."

temperature changes, but the water itself is the very structure of the fish's space. A fish can't imagine (if a fish could imagine) that the surface of the water marks the end of its universe any more than we can contemplate an edge to our universe, a beginning or an end to time. Fish evolved in the backdrop of water, just as human beings evolved in the equally insensible backdrop of our own particular vacuum. Is it any wonder that nothing is unthinkable? Or that when we do try to think about it, we're like fish out of water?

People well trained in the art and science of Zen meditation may be able to "sense" nothing. But to think about it—to analyze its characteristics—requires that kind of structured thought that by definition destroys the very nothingness you seek. Partly for this reason, most physicists don't even try to think about something as amorphous as nothing. (There are exceptions.) Once you get close to it, nothing evaporates, because there is no "it" for it "not" to be.

Nothing can also be hard to grasp because it so easily masquerades as something, and vice versa. The motion of a car cruising along at 70 mph seems like nothing if you're sitting inside—until it hits a wall; air pretends to be nothing until the wind whips it into a hurricane and rips trees up by their roots, or a spark ignites it, or it holds a 500,000-pound jetliner aloft; gravity can't be touched with our hands or seen with our eyes, but it glues us to the ground just the same; radioactivity remains invisible even though it can—and sometimes does—kill us. We don't notice most of the things that are done and said right in front of our noses.

Indeed, almost any kind of nothing you can conceive is actually not. In the emptiest corners of interstellar space, a spoonful of vacuum might contain an atom or two. Even if you could somehow get rid of all the matter, you'd still have light streaming in from billions of sparkling star-strewn galaxies—not to mention all those even more exotic sparklers beaming out X rays and gamma rays and gravity waves and neutrinos and heaven knows what else.

Even if you could get rid of all that, you'd still have the *vacuum*—which has both form and content.

"The vacuum means that nothing more can be removed," says the man who knows more about nothing than almost anyone, Harvard physicist Sidney Coleman. "It's an empty box. But that doesn't mean that the empty box has no structure."*

And then there is space and time. "Do not underrate the importance of time and space," writes Einstein scholar Banesh Hoffmann in *Relativity and Its Roots.* "They may seem to be intangible nothings, less palpable even than the faintest breeze."

But space and time, he points out, have immensely powerful properties. When we run away from danger, we use space as a shield. "Certainly if space can thus protect us, it is not a soft nothingness. Nor is time. How safe we would be from death by nuclear bomb had we been born in the time of Shakespeare."

Woven together, space and time are the warp and woof in the fabric of four-dimensional spacetime, an entity that can curve and wiggle and swallow stars whole. Objects created from nothing but spacetime include almost certainly real things like black holes, possibly real things like wormholes, and everyday things like gravity.

In physics, "nothing" has very specific properties—not all of them obviously compatible, some leading to more curious dualities. These apparent contradictions in the nature of physical nothing may well be the cracks that open the door to understanding how the universe evolved.

For example, one working definition of nothing is "that which doesn't make a difference." Nothing is what you can't tell you're in

---

*See Chapter 4, "Nothing Takes Center Stage."

because it doesn't look different whether you're going up or down, left or right; whether you make it hotter or colder; whether you travel forward or backward in time; whether you're moving at light speed, or not at all. No matter what you do to nothing, or in nothing, it always looks the same. It's an endless cloud of undifferentiated gauze that obscures all differences, a place where you could easily get lost, no way to tell in from out, or up from down, or today from yesterday.

You can't change nothing by definition because nothing you do to it would make a difference. This immutability makes nothing the most fundamental stuff in the universe.

When nothing changes, we know it right away. When nothing changed, the universe was born.

Something, by this definition, is any deviation from nothing. Nothing is the norm; something is derivative. We create something by breaking the perfect symmetry of nothing, cracking the silence—like drawing black lines on a white piece of paper, or introducing ripples into perfectly still water or kinks into energy fields.

This way of thinking suggests that nothing is perfection—or at least, perfect symmetry, which to many physicists is the same thing. Nothing is perfect, but not very interesting.

Of course, this notion of nothing leads smack into a tautology: If nothing is by definition undetectable, then you can only prove it exists by its absence.

Physicists can't deal with this nothing because there's nothing to be seen. Besides, the nothing behind the physical universe—which is, after all, the real subject of physics—is not a perfectly transparent perfection. Rather, it is a shattered perfection, like shattered glass. The cracks in perfection allow it to be studied.

The nothing that concerns physicists is what you have left after you remove everything you can possibly take away. Nothing, in other words, is the state of lowest possible energy. This kind of nothing isn't perfectly symmetrical. It has an underlying structure—like Coleman's empty box. But you can't suck anything else out of this nothing.

Say the universe is made of vibrating strings; if you take away all the energy, the strings don't vibrate, but you still have strings. Or let's say "nothing" is the watery universe of a fish. If all the energy is taken out, the water freezes. Instead of an amorphous fluid, you now have the crystal structure left behind when all the energy is removed. But it's still water. This structured kind of nothing corresponds to the vacuum in which we (and everything in the universe) ultimately reside.

The evolution of the universe, roughly speaking, started with a perfectly symmetrical nothing, which cooled into our current frozen vacuum state and may one day melt down again.

The first kind of nothing, according to the reigning "pope of string theory," Edward Witten, embodies the laws of nature in their ideal, perfectly symmetrical state. The second nothing reflects the laws as they appear in the universe we actually live in.

Physicists seeking the ultimate laws of nature search for perfect symmetry—the ultimate, perfect nothing. But to understand it, they first have to understand the colder, everyday nothing we live in—and the various kinds of vacua involved in giving it birth.

NOTHING SPECIAL

Nothingness haunts being.

—JEAN-PAUL SARTRE, *Being and Nothingness*

So, baffling or not, we need nothing. Among other reasons, we need it to create something, anything. Asentencelikethisdoesnotmakeawholelotofsenseevenifyouunderstandallthewords. Empty spaces put borders around things, picking them out from the crowd. As Lucretius argued two thousand years ago in his classic, *De Rerum Natura* (still a source of great insights about everything from physics to love, by the way), things can't move if there aren't empty spaces for them to move *into.* Without a vacancy, there is nothing to fill.

People who remember everything experience mental confusion as severe as that of people who remember nothing. There is no music or speech without silence, no life without sleep.

"Although space may be empty of body, nevertheless it is not in itself a void," wrote Isaac Newton. "*Something* is there, because spaces are there, although nothing more than that."

If nothing else, nothing is a crutch, a creation of convenience that allows us to perceive and understand the nature of something, a breathing space that makes the sometimes overwhelming richness of the universe manageable.

Modern physics tells us—contrary to Lear—that you *can* get something from nothing. But it's also clear that you can't get something *without* nothing. Nor can you have nothing without something.

Sartre, whose *Being and Nothingness* goes on endlessly about the subject, makes it clear that the idea of nothing hinges explicitly on the idea of something: "Nothingness, which is not, can have only a borrowed existence, and gets its being from being. . . . [T]he total disappearance of being would not be the advent of the reign of non-being, but on the contrary the concomitant disappearance of nothingness."

Nothing is not only indispensable. It is unique. That's why absolute nothings occupy a special place in physics. The temperature absolute zero is not at all the same as *nearly* absolute zero; for one thing, the former is unreachable. Absolute zero resistance produces strange behavior that is not even suggested at *almost* zero resistance. For example, liquid helium flows up and out of bottles and seeps through the bottoms of ceramic containers.

THE BEGINNING OF NOTHING

The old Church fathers did everything to keep it out of a world which then revolved around one and its multiples. . . . Zero was unthinkable. If it wasn't one of something, it couldn't be allowed.

—SOPHIE PLANT, *Zeros + Ones.*

The obvious need for nothing led resourceful peoples all over the world to invent it. Nothing is a human creation, like Beanie Babies, or wheels. The mathematical nothing—the early ancestor to zero—arose in at least several different cultures, only to be beaten down repeatedly by the skeptical, the unwilling, the repressed.

In a way, the skeptics were right. Once let out of the closet, zero began acting out like some wild adolescent, procreating entire new branches of the mathematical family tree. Along with its evil twin, infinity, zero set generations of mathematicians and philosophers to pondering its elusive properties. But like children, zero also produced enormous and equally unexpected pleasures. Probably the most powerful applied tool of mathematics—the calculus—is based on the idea of approaching zero while never actually touching it.

Much the same can be said of Lucretius's notion of empty space as a simple holding pattern, a blank page on which anything can be written. It, too, almost immediately began to take on all sorts of troubling properties.

In a way, the history of physics is a history of turning nothing into something—and vice versa. We used to think matter was solid, material. Now, we know that atoms—and everything made of them—are mostly empty space. What's more, that empty space is a prolific producer of particles and energy fields of all sorts.

One of the most notorious crises in this on-and-off relationship between physics and nothing revolved around the luminiferous ether, the stuff of empty space before Einstein came along and showed that the ether was undetectable because it didn't exist. Einstein cleaned up the previously accumulated clutter in the vacuum with the broom of relativity. But at the same time, he changed empty space from a staid background set into an active player that gives as well as it gets.

Meanwhile, the new understanding of the atom revealed that particles of matter are chunks of vibrating fields, condensations of energy into a kind of frozen state. The particles acquire the sluggishness associated with mass by interacting with something called the *Higgs field*—the twentieth-century ether. But the Higgs field itself is part and parcel of our vacuum, and physicists like Frank Wilczek of the Institute for Advanced Study call the long-sought Higgs particle a "chip off the old vacuum."

This modern version of the vacuum is once again weighed down with too much unnecessary stuff, complains physicist Leon Lederman. "[T]he pristine absoluteness of the vacuum state (as a concept) has been so polluted (wait until the Sierra Club finds out!) by twentieth-century theorists that it is vastly more complicated than the discarded nineteenth-century aether."

Not only has our notion of nothing evolved, so has the modern vacuum itself. If the physicists are right, our universe does not reside in the same kind of vacuum it did at the beginning of time. Indeed, according to current theories, every speck of matter and energy in the universe emerged from a now mostly disintegrated "false vacuum." Our current vacuum is a "frozen" state of that previous vacuum state.

So even nothing itself evolves.

In a sense, nothing gets created in new forms as it becomes better understood, more clearly defined. Just as species and children evolve (sometimes bizarrely) under the influence of chance and necessity, so has the scientific concept of nothing.

NOTHING: A WORK IN PROGRESS

String theory is telling us that space and time are not like anything we are used to. But please don't ask me what it will be, because I wish I knew.

—physicist NATHAN SEIBERG, Institute for Advanced Study

Truth be told, physics still hasn't completely recovered from the revolution that Einstein wrought. Indeed, some would say that this revolution remains largely unfinished. When the dust clears, "nothing" may well be unrecognizable once again.

After all, all physics takes place in a context of space and time. Space and time are as close to nothing as physics usually gets. Once you start meddling with space and time—as Einstein did—all bets are off.

"So familiar are time and space that we are apt to take them for granted, forgetting that ideas of space and time are part of the shaky foundation on which is balanced the whole intricate and beautiful structure of scientific theory and philosophical thought," writes Banesh Hoffmann. "To tamper with those ideas is to send a shudder from one end of the structure to another."

Einstein did just that. Einstein turned solid spacetime into a soft mush, as malleable as Silly Putty. But even then, it still served as an ultimately indestructible background for everything else. Now spacetime itself may be in danger. In certain versions of string theory, at least, "space and time are doomed," according to Witten.

One of the major fruits of string theory is that it has brought questions of the true nature of space and time back to the forefront of physics. It's not even clear whether the "strings" exist within space and time, or whether space and time somehow emerge from the strings. Columbia University physicist Brian Greene, for example, tentatively describes the strings as "shards" of space and time that somehow assemble on a larger scale into the seconds and miles in which we measure our everyday lives.

At the almost unbelievably tiny scale of strings, space and time melt down, dissolve into something—for now at least—completely ungraspable. "You're in a spaceless, timeless realm in those submicroscopic depths," says Greene.

Space and time may turn out to be superfluous in a universe made of strings. If so, then our present-day image of the fabric of spacetime may unravel completely at some fundamental level, making this notion as quaint and outdated as the ether.

"We're getting closer and closer to a major revolution," says Seiberg. "Things are going to get wilder and wilder."

One thing that's all but certain, however, is that space itself has expanded into several new dimensions—literally. Where Einstein's revolution replaced everyday notions of three-dimensional space with four-dimensional spacetime, string theory takes place in a complex geometry of at least eleven dimensions of space and time.

If space is nothing, there's far more of it to explore in modern physics than there's ever been before. "Space is undergoing a drastic rearrangement of its pieces," says Greene.

And at the same time that string theory is overhauling spacetime, it's also once again propelled the mysteries of the vacuum to the fore. String theory, it turns out, seems to come equipped with far too many vacua, and no one knows exactly what that means.

Usually, physicists have little problem finding the lowest energy state of whatever they happen to study: a rock at the bottom of the hill, a string that isn't vibrating, or a hunk of matter at its lowest possible temperature.

String theory, however, seems to produce an infinite number of lowest energy states.* That is, it produces an infinite number of vacua. No one knows how to decide which vacuum is the right one—the one which represents our universe—or why.

---

*See Chapter 6, "Nothing Gets Strung Out."

String theory aside, modern cosmology is also drastically changing the nature of nothing: bluntly put, there seems to be more and more nothing all the time—even in the traditional four dimensions of spacetime that we know about. The universe is expanding because nothing is continually being created in the spaces between clumps of matter. This nothing—this vacuum—appears to have enough energy to push the galaxies at the far reaches of the universe farther and farther apart.

NOTHING BECKONS

If you get a Big Bang or Genesis right, the structure of the subsequent physical and historical world will make sense.
—physicist MARTIN KRIEGER, *Doing Physics*

Clearly, if nothing is behind everything, it behooves us to understand what it is—or at least to explore its dubious properties. Every account of the origins of the universe—whether it comes from physics, or from religion—begins (and sometimes ends) with a void.

"Whether it be Nature or Scripture, it turns out that we face the same problem in describing how an orderly world of something arises out of a soup of disorder and chaos and nothingness," writes Krieger. "The obsession with creating an orderly vacuum or an orderly social world is a generic one about origins and creation."

What, then, is the nature of this parental void, our common ancestor? What is its shape, form, weight, color, stiffness, smell? If we can figure that out, we can be privy to the ultimate secrets, the underlying laws. Like it or not, we are probably destined to spend eternity desperately seeking nothing, like Sisyphus pushing his stone uphill.

In a way, our task is more difficult, because our goal is to understand an abstraction, and abstractions always weigh heavily on the mind. It's easier coming to terms with a concrete curb than stumbling over such concepts as the vacuum or curved spacetime. Not to mention nothing at all. Is nothing truly real enough to take seriously?

The truth is, almost every solid idea that comes from science is in some sense an abstraction rather than a "real" thing. An atom is not something you can put your finger on. Gravity is "seen" only by its insistently attractive behavior. Light is a traveling kink in an electromagnetic field. And yet, light and gravity and atoms are the very substance of our world.

A copy editor at the *Los Angeles Times* once asked me how anyone could say for sure that spacetime is really curved. "You can't see it," she argued.

But we don't see atoms and we don't see air, and they affect us just the same.*

Perfect circles and squares can't exist in the real world, either, but that doesn't stop generations of geometry teachers from introducing them to schoolchildren as concrete objects with real properties.

Indeed, the things we see with our eyes and brains—even "concrete" objects such as curbs—are largely created in the mind's eye as, yes, abstract concepts. We see by reassembling the muddled bits of information that fall on the backs of our eyes into sensible images. Then we "project" the images inside our heads onto the world at large, just as we project the sounds we hear inside our heads to outside sources.

The reality of these perceptions—the concreteness, if you will—is well known to neuroscientists: a person with a missing arm or foot may perceive very real pain or pleasure in that absent limb. Anyone who has felt the chill wind blowing through a hole in the heart knows just how tangible an abstraction can be.

---

*And, of course, we *can* see curved space. See Chapter 5, "Nothing *Becomes* Center Stage."

Nothing, in this context, is every bit as real as a circle, an atom, gravity, a perception, a chair. It just seems strange because it is unfamiliar. But, as the late physicist Sir Arthur Eddington pointed out in an explanation of Einstein's four-dimensional spacetime, believing strange—even impossible—thingsis a requirement for understanding the workings of our universe.

In Eddington's still wise words:

> It is difficult to ignore a voice inside us which whispers "At the back of your mind, you know that a fourth dimension is all nonsense." I fancy that that voice must often have had a busy time in the past history of physics. What nonsense to say that this solid table on which I am writing is a collection of electrons moving with prodigious speeds in empty spaces, which relatively to electronic dimensions are as wide as the spaces between planets in the solar system! What nonsense to say that the thin air is trying to crush my body with a load of 14 lbs. To the square inch! What nonsense that the starcluster, which I see through the telescope obviously there *now,* is a glimpse into a past age 50,000 years ago! Let us not be beguiled by this voice. It is discredited.

We create the concepts that allow us to perceive the world. Call them abstractions if you like. But they are as real as things get. "When we say that a thing is real we are simply expressing a sort of respect," says physicist Steven Weinberg.

Nothing deserves, at the very least, our respect.

Chapter 2

NOTHING HAPPENED

We calculate; but that we may calculate,
we had to make fiction first.
—NIETZSCHE

IN THE BEGINNING, there was nothing. Not even zero. Not even empty space. Not even the vacuum. Human beings stumbled upon zero and nothing by accident—and recoiled in horror. Feared, reviled, and sometimes banned outright as negative (even unholy) influences, these proverbial ugly ducklings would have to wait many centuries before flowering into their full potential; certainly before becoming recognized as "the turning-point in a development without which the progress of modern science, industry, or commerce is inconceivable," in the words of mathematician Tobias Dantzig.

The evolution from despised misfit to favorite child was neither direct nor smooth. Both zero and nothing came into existence as mere placeholders, pauses between one thing and another—like a dash between words. They were passive nonentities, easy to ignore.

Only when people tried to pin them down did they begin to ooze out at the edges, expanding into entire fields of study, giving birth to number systems and universes. Like living species, their evolution was directed by the pressure to adapt to constantly changing intellectual environments, as well as much random swapping of (cultural) genes.

For in any kind of evolution, as much is owed to luck as to purpose or even creativity. "A great discovery! Yes," Dantzig goes on to say about the invention of zero. "But, like so many other early discoveries . . . not the reward of painstaking research, but a gift from blind chance."

The evolution continues to this day. And the spurts and starts and near disasters tell us as much about the mindset of the people who continually reconceive nothing as the meaning of nothing itself. The sum total of their efforts, however, has changed nothing from an absence of everything to the existence of something unlike anything the world had ever seen.

THE STORY OF 0: HOW NOTHING BECAME SOMETHING

In the history of culture the discovery of zero will always stand out as one of the greatest single achievements of the human race.

—TOBIAS DANTZIG, *Number: The Language of Science*

In Russian, the word for zero is *nool;* in Spanish, it's *cero;* in Arabic, it's *ṣifr*; in Hungarian, *nulla*; in Japanese, *rei*; in Esperanto, *nul*; in Klingon, *pagh.*

A Buddhist by birth, the modern English zero derives ultimately from the Indian *śūnya,* or "void," that fertile emptiness that is the essence of all things, the primordial egg. From Hindu through Arabic through Latin, Italian, English, Esperanto, Klingon, its names trail its evolution like crumbs of bread left behind to mark the way.

As a symbol, zero's meaning is far more blatant: a little bit of empty space corralled inside a circle, captured, locked up. You might say the symbol zero is a container for a little bit of nothing.

The words and the symbol came late on the scene, well after all the other numbers, an afterthought. When counting, people put it first, but in order of invention and acceptance, it came last—by many centuries, in fact.

It's difficult to pin down zero's exact birth date, in part, because it was invented and reinvented in many different forms by different civilizations. People argue over whether it first appeared in India or among the Mayans or whether the Arabs were really responsible for anointing it as a number. These squabbles are beside the point.

"The story of zero typifies much of the history of mathematics," says William Dunham in *The Mathematical Universe.* "An idea is born; it is refined and transmitted over the miles and over the centuries; it becomes a part of the multinational mathematical culture. Mathematics is a creation in which all the world can proudly share. Or so it should be."

In fact, a case can be made—and is, quite compellingly, by Robert Kaplan in *The Nothing That Is*—that "every ecological niche in the world of zero's possible origins has been filled."

This much is certain: Zero was Greek, as the saying goes, to the Greeks. Neither Pythagoras with his theorem nor Euclid in his element thought of it.

The ancient Babylonians, however, seemed to have developed a way of noting an absence of a number by inserting a kind of wedge-shaped marker in the place where it failed to appear. This early proto-zero, first seen (but not consistently) around 500 B.C., was the absence of something, rather than the presence of nothing. Over time, the wedge became a dot, then a small circle, then the familiar zero.

By the first century A.D., zero was definitely in use by both the

Mayans and the Indians. The Mayans represented zero with the figure of a head of a man, hands clasped around his jaws. The Indian zero, according to Graham Flegg in *Numbers: Their History and Meaning,* was largely an invention of Indian astronomers, who needed a way to keep track of very large numbers—easy enough to do when adding a few zeros expands quantities a thousandfold.

Somewhere around the ninth century, the Indian notation spread into the Arab world, where *śūnya* became *ṣifr.*

The Western world really didn't play much of a role in zero's career until a young Italian named Leonardo Pisano, schooled by Muslim teachers in the Hindu Arabic number system, took it upon himself to bring these new numbers to the West—still mired at the time in clunky Roman numerals. Those ponderous Xs and Ms and Ls were impossible to put to use for calculation; one could only line them up like so many classical pillars, set in concrete (or marble, as the case may be). They had no sense of place; an X at the beginning of a sequence could mean almost the same as an X at the end; the only way to add was to pile on more letters, or invent new ones. Roman numerals were about as useful for math as hood ornaments. They could only make a statement, not take you for a ride.

Known more famously today by his nickname, Fibonacci, Pisano described our now familiar number system in a book called *Liber Abaci,* published in 1202. But even then, zero did not stand on equal status with the other numbers. Fibonacci referred to the numbers one through nine as "figures," but zero only as a "sign," like plus or minus. (It was promptly denounced as "heathen.")

Hundreds more years would pass until zero popped up as an actual quantity that could be added, subtracted, multiplied—that is, until it became a real number. Even then, it could only be used in arithmetic and geometry. Zero could designate no quantity, no dimension, no angle, no side, no length, no width. But that's about all. Its more curious permutations would have to await the invention of algebra, and all the rest.

NOTHING IN THE WAY

The zero is something that must be there in order to say that nothing is there.
—KARL MENNINGER, *Number Words and Number Symbols: A Cultural History of Numbers*

Zero's lowly origins remind one of those spacers used to keep children's teeth apart while they're waiting to get fitted for the first set of what are now called orthodontic "appliances." Its only purpose was to leave a space where something else might grow. It wasn't an object in its own right any more than lack of intelligence is a sign of intelligence, or absence of trees qualifies as a variety of tree. It was merely a poor stand-in, a mannequin that saves a place in line or a strut that holds beams apart.

This kind of placeholder zero didn't translate well into writing. Picture a counting system based on pebbles in the sand or marks in clay or beads strung on an abacus. It's easy enough to see when a pebble or mark or bead is missing. But when you start to translate the objects into some kind of symbolic notation, problems immediately appear. You can leave a space, but how big should it be? What if there are two empty slots? How do you designate the difference? How big a blank should there be?

The transition from simply leaving an empty space to filling the space with a symbol marked a huge step in the evolution of zero. In short, it marked the transition from the absence of something into the presence of a real nothing. Instead of simply saying: "Nobody home," zero now signified "one (or two) missing object(s) here."

This was no mean feat, in that it allowed the concept of nothing to exist aside from the lack of any particular thing: no pebbles, no beads. Zero helped move arithmetic off the abacus and onto slates and paper. "Zero," writes Menninger, "liberated the digits from the counting board and enabled them to stand alone."

In giving the concept of nothing life on the page, zero allowed it to begin its slow evolution into something entirely new.

We're so accustomed to thinking of zero as a number that it's hard to remember just how odd that idea really is. Think of a guest who doesn't show up for dinner. Do you treat that guest the same as all the others? As an actual, present, person? Or what about the money you don't have? What can you buy with it? Or the language you never learned. Is it Latin? Or French?

Still, even as a punctuation mark, a symbol for nothing serves a critical purpose. It provides a pivot where the worm—or the meaning—can turn. As the writer Dick Teresi points out, spacecraft counting down before blastoff use the count of "zero" as the marker between one phase of flight and another. Counting down to zero, spaceflight is controlled from Cape Canaveral in Florida. Counting back *up* from zero, the flight is handled by Houston. Zero is the pause that allows the transition to occur.

At first, this placeholder zero was used only in the middle of numerical sequences. It could spell the difference between 4005 and 45, but not 45 and 4500.

When people started putting zero at the end of a number, they literally multiplied that number's value tenfold. Now zero's position, as well as its value, counted for something. Like the silent *e* at the end of a word, an otherwise worthless zero at the end of 45 could change something's value completely, even though the zero had no value in itself. As such, it became the wind beneath the wings of the other numbers, allowing them to soar to previously unattainable heights.

Clearly, these developments took place in different guises in various cultures around the world and at various times. For example, a kind of "zero" also appeared in an Incan counting system in the fifteenth and sixteenth centuries. Instead of a counting board or abacus, the Incas encoded numerical messages on collections of knotted multicolored cords called quipus. In general, the absence of a knot on a cord meant an empty space. But there were subtle distinctions. A blue cord with no knots, according to Marcia Ascher in *Ethnomathematics,* would indicate that zero potatoes were eaten by children in a family; the absence of a blue cord, on the other hand, meant the family had no children.

NAUGHTY, NOT NICE

As usual, prohibition did not succeed in abolishing,
but merely served to spread bootlegging
—TOBIAS DANTZIG, *Number: The Language of Science*

One aspect of zero's history that was nearly universal, however, is the chilly reception it got almost everywhere it showed up. Small wonder. What's the point, one might well ask, of counting nothing? "Zero was a long time in establishing itself as a number in its own right," says Jan Gullberg in *Mathematics: From the Birth of Numbers.* "It was hard for mathematicians to accept that 'nothing' could be regarded as 'something.'"

The Greeks certainly couldn't bring themselves to do it, possibly because—as some have suggested—they were too deeply cloaked (perhaps even suffocated) by their purist philosophy. A worldview constructed around the clean, clear lines of perfect geometric forms simply couldn't allow for something as heretical as a nothing of substance. Perhaps, as some present-day executive might put it, the Greeks simply couldn't bring themselves to think "outside the box."

"The concrete mind of the ancient Greeks could not conceive the void as a number, let alone endow the void with a symbol," speculates Dantzig. In contrast, "the Hindus were not hampered by the compunctions of rigor, they had no sophists to paralyze the flight of their creative imagination."

Other writers are somewhat more generous in excusing the Greeks. After all, Greeks didn't need zero. Their forte was geometry: numbers were forms; positive integers suffice for that. They had no need of lines of zero length, angles of zero degrees, or circles of zero radius.

During the Middle Ages, zero's existence was acknowledged, but surely not welcomed. Indeed, it was "regarded as the creation of the devil," says Menninger—although this surely tells us less about zero and its Hindu-Arabic companions than the countries the infidel number traversed as it slowly made its way from East to West. Indeed, edicts were issued in Florence and elsewhere that explicitly forbade the use of the new number system.

Others merely scoffed, putting down the zero as a puffed-up pretender to the status of number. Sniffed one unidentified fifthteenth-century source: "Just as the rag doll wanted to be an eagle, the donkey a lion, and the monkey a queen, the *cifra* [zero] put on airs and pretended to be a digit."

To call someone a zero was something like calling a person "blockhead" today. And even now, few people take being called a "zero" as a compliment. The word *naughty,* after all, comes from another term for zero—*naught.* So does *cipher,* which not only tars someone as completely insubstantial, but speaks of zero's slightly seedy, undercover origins.

It's easy to see why zero caused so much trouble. Simply counting (never mind higher math) becomes immediately confusing. Consider that, even today, people can't seem to agree on whether the ground floor of a building is the first floor or the zero floor. Odometers start with zero, but birthdays with one; temperature begins with zero degrees, but school with first grade.

And these problems only hint at the trouble to come when people started to treat zero on equal footing with its fellow digits, attempting to add, subtract, multiply, and—worst of all—divide. (See Chapter 3.)

A VERY BRIEF HISTORY OF EMPTY SPACE

Then even nothingness was not, nor existence.
—*The Rigveda,* twelfth-century B.C. Hindu creation hymn

The vacuum is a garbage dump. Einstein freed us from it, now we've got to get rid of it again. Some kid now in junior high school will tell us how.
—LEON LEDERMAN, Nobel laureate in physics

Compared to the coming of age of empty space, the early years of zero were downright tame. Space began, like zero, in total obscurity. To the earliest human thinkers, the idea of nothing was unfathomable. Outside the sealed celestial dome of the stars, "there was nothing, not space, not even emptiness," as Daniel Boorstin reminds us in *The Discoverers.*

But once nothing came out, it set off a series of heated controversies that continue unabated until this day. Over a period of several thousand years, it went from a nonentity to a background, or stage, and then became the stage itself: a stage that moved, acted in the play, created its own characters, and generally took over as producer, director, set designer. As if all that wasn't enough, it began to expand, spinning off progeny in the most prodigious production of "new spaces" known, well, in our (and other) universes.

Like zero, though, empty space began as a simple placeholder, a space to put things that had nowhere else to go, an ingredient of nature necessary to keep things from falling in on themselves.

"If there were no empty space," Lucretius wrote, sensibly enough, "everything would be one solid mass."

If you believed in atoms, nothing was a necessity. In order to move, atoms, and objects composed of them, needed a place to move *into* when they changed places. Without the void, there would be no elbowroom for motion. The material universe would be like a game of musical chairs with no open seats and no place left for anyone to go.

No wonder Lucretius insisted: "*There is vacuity in things*" (emphasis his) in his popular poem about physics, love, and philosophy, published in the first century B.C. "By vacuity, I mean intangible and empty space. If it did not exist, things could not move at all. . . . Nothing could move forward, because nothing would give it a starting point by receding."

This vacuity, as Lucretius described it, also accounted for such properties of things as heaviness or lightness of being. Air weighed less than water, which in turn weighed less than rocks, because air had more vacuity packed into it than water; water, in turn, was more vacuous than rock.

Lucretius's *De Rerum Natura* served as a compelling argument for the theory of matter advanced by Democritus hundreds of years earlier. In 420 B.C., Democritus said that everything in the universe was composed of atoms and the void. By giving empty space half the billing, he gave it a central—if not starring—role in the cosmos.

Democritus's views were not popular, however, and remained virtually unknown—or outright banned—until the Renaissance.

In fact, the idea of nothing inspired absolute terror in Greeks such as Aristotle. "The prospect of an unclassifiable emptiness, an attributeless hole in the natural fabric of being," writes Brian Rotman in *Signifying Nothing,* "presented itself as a dangerous sickness, a God-denying madness that left him [Aristotle] with an ineradicable *horror vacui.*"

MUCH ADO ABOUT A NONENTITY

{Empty space} sways irregularly in every direction as it is shaken by those things, and being set in motion it in turn shakes them.

—PLATO, on empty space, from his dialogue, *Timaeus,* quoted in Nick Huggett's *Space from Zeno to Einstein*

Plato, who influenced a whole line of empty-space skeptics—including Aristotle, Descartes, and Leibniz—nevertheless had a great deal to say about its properties. Space, he said, is the essential matter out of which all material things are made. Material objects are regions of space sculpted into the form of matter—cups, cows, olives. As Plato wrote of space in his famous dialogue *Timaeus*: "Its nature is to be available for [any element] to make its impression upon, and it is modified, shaped, and reshaped by the things that enter it."

What's more, in language that almost sounds like Einstein's overhaul of spacetime in the early part of this century, Plato described how space not only creates matter, but "shakes" matter as well. Matter, in turn, shakes up empty space.

As we'll see later, this isn't so far from physicist John Wheeler's famous summing up of Einstein's spacetime: "Matter tells space how to curve, and space tells matter how to move."

Even Aristotle, for all his *horror vacui,* paints a surprisingly active picture of what we might today call empty space (even though he didn't). Because space is "place" in Aristotle's world, it accounts for the motion of objects. That is, heavy things fall because they are in the wrong place and are seeking the right one. Light things rise because they, too, are seeking their proper, elevated, place in the cosmos.

So space is a "cause" of motion, a notion that harkens ahead to the potential packed inside otherwise empty space—potential that can lead to the creation of particles or universes. In another thoroughly modern take on space, Aristotle wondered whether space could grow, expanding along with any object that grew, increasing its territorial hold, and generally taking over more and more of the universe. He concluded that the question was moot—because space was not really, he said, distinct from matter—but his entertainment of the possibility remains prescient nonetheless.

Space was expanded for real by the fifteenth-century philosopher Nicholas of Cusa, the first person to seriously think that space might extend all the way to infinity. Although he based his ideas on theology, not physics (Cusa was a Roman Catholic cardinal), Cusa nevertheless threw open the doors of the closed ancient cosmos, allowing a larger perspective. Arguing that space could have no center, Cusa also concluded it could have no edge and therefore no end.

Many other thinkers took up the issue of the nature of emptiness—not always with perfectly consistent results. For example, in the seventeenth century, Descartes followed Plato by arguing that empty space was a nonentity, since it couldn't possibly be measured. (How could you attach a measuring device to nothing?) However, Descartes did believe that the heavens were filled with "liquid matter," which not only pushed planets around but also got sucked into vortices that condensed into palpable objects.

(Descartes, like Aristotle, could give nothing such palpable properties because he called nothing a "substance." More on that later.)

SPACE HEATS UP

{Absolute space is} a conceptual monstrosity, a purely thought-thing which cannot be pointed to in experience.

—Ernst Mach, quoted in Gerald Holton's *Einstein, History, and Other Passions*

The whole debate heated up again when Newton made space a central part of his mechanical universe. Newton's space was passive. It could neither affect matter nor be affected by matter. But it provided an essential framework for everything that happened in the cosmos. Completely independent and objective, empty space was absolute. Consider it a series of grid lines that mark off the warp and woof of the universe, a fixed system of coordinates that pinpoint the position of every object that exists, every event that happens. While it's not directly visible, to Newton, it was out there—an essential ingredient of the physical world, as real as mass or motion or time.

Others said, "Bah!" Most notoriously, the German philosopher and mathematician Gottfried Wilhelm Leibniz insisted that Newton's space and time were mere illusions. Space—as well as time—was only a way of thinking about things, a way of perceiving relationships between clumps of matter and events.

Leibniz compared space to a genealogy, like a family tree. While it's certainly helpful in delineating relationships, it does not exist as a concrete object. There's no tree there, and no space there, either, according to Leibniz. Granting the status of "real" to Newton's coordinates in space was as silly as thinking that the bloodlines which run in families are really lines. They are only, Leibniz insisted, "Ideal Things," not reality.

Nevertheless, Newton's view prevailed. Indeed, most people today still believe in their heart of hearts that absolute space and time are somehow "out there," fixed and immortal, serving as guide stars that keep us from going astray as we grope through the universe. These people are going to have to hang on to their hats in all that follows.

Suffice it to say for now that Newton's sedate and silent backdrop was shattered as soon as electricity and magnetism began to be understood as dynamic fields of influence (not unlike political spheres of influence) that pervade empty space. The fields took on lives of their own, creating and sustaining everything and nothing alike. No wonder Einstein called the concept of field "the most important invention since Newton's time." As he and Leopold Infeld remark in

The Evolution of Physics, "It needed great scientific imagination to realize that it is not the charges nor the particles but the field in the space between the charges and the particles which is essential for the description of physical phenomena."

Einstein, of course, went on to do more to and with nothing than all his predecessors combined. It's clear he held the subject foremost in his mind from the earliest days. In 1902, he drew a sketch of his apartment in Bern, and sent it to his wife-to-be, Mileva. In charming detail, he designates the exact placement of every piece of furniture in the room—lamps, dressers, and chairs—and even of the doors and windows.

In the center, though, a symbol floats eerily apart from everything else, in the empty space where nothing is. Einstein couldn't leave well enough alone. He drew in the symbol for what would soon be the center of his professional life: Nothing.*

THE E---R PROBLEM: A SPECIAL CASE

Quite undeservedly, the ether has acquired a bad name.
—FRANK WILCZEK, in *Physics Today,* January 1999

One epoch in the early history of empty space deserves special treatment because it probably caused more trouble than any other incarnation of nothing. Indeed, this so-called luminiferous ether gave physics such a black eye that it's often invoked as an example of the way physicists sometimes seduce themselves into believing the inventions of their own fertile minds, even when nature provides ample clues to the contrary.

Einstein himself called the ether the "*enfant terrible*" of physical substances. "After such bad experiences," he wrote, "this is the moment to forget the ether completely and to try never to mention its

*The drawing is reproduced in Holton's book, *Einstein, History, and Other Passions* (Reading, Mass: Addison-Wesley, 1996).

name." In his further discussion, he refers only to "the 'e---r' problem"—as if the very word qualified as profanity.

So elusive that every attempt to pin it down evaporated in frustration, the ether nevertheless settled like concrete into the minds of physicists for many centuries—even millennia. In the nineteenth century, Lord Kelvin spoke for many of his colleagues when he pronounced: "One thing we are sure of, and that is the reality and substantiality of the luminiferous ether."

But the ether may well be, as Wilczek suggests, merely misunderstood. As even Einstein pointed out, "This word ether has changed its meaning many times in the development of science."

Today some physicists believe the most recent reincarnation of the ether may account for as much as 70 percent of the energy in the universe.

ANCIENT ETHERS

Perhaps the whole frame of nature may be nothing but various contextures of some certain etherial spirits or vapors, condensed as it were by precipitation. . . . Thus perhaps may all things be originated from ether.

—ISAAC NEWTON, letter to Henry Oldenburg, 1675

In one form or another, the ether has been around for a long time. It came, originally, straight from heaven.

Aristotle ascribed wondrous qualities to the ether, which he called the "fifth essence," or *quintessence.* (The first four essences, or elements—earth, fire, air, water—made up everything on earth.) Transparent, weightless, and generally otherworldly, the ether was a kind of heavenly mist that made up all matter and form in the higher reaches of the cosmos, the pure breath that pervaded the playing fields of the lords. Divine stuff, this ether. Not for us. Not in our backyards. Like ambrosia, it was reserved only for inhabitants of heaven.

In a sense, Aristotle could dispense with nothing because he had the ether to fill the void. Ether accounted, for example, for the circular motions of planets and stars. The "natural motion" of the ether, according to Aristotle, was circular. So stars and planets twirling through the ether were merely (in another echo of Einstein) following their natural paths in space (or ether).

Descartes, another sworn enemy of the vacuum, also relied on the ether to push the stars (and planets) around. Vortices of swirling ether stirred up the heavens like a van Gogh sky, condensing into objects and generally providing the unseen hand behind motions and matter alike.

Newton himself was a firm believer. In fact, in a letter written in 1675, he seems to base just about everything in nature—not only what *is,* but also what *happens*—on the ether:

> [I]t is to be supposed that the ether is a vibrating medium like air, only the vibrations [are] far more swift and minute. . . . Now these vibrations . . . may be supposed the chief means by which the parts of fermenting or putrifying [*sic*] substances, fluid liquors, or melted, burning, or other hot bodies, continue in motion, are shaken asunder like a ship by waves and dissipated into vapors, exhalations, or smoke. . . .

GRASPING FOR ETHER

All assumptions concerning the ether led nowhere!

—ALBERT EINSTEIN and LEOPOLD INFELD, *The Evolution of Physics*

So what *was* the ether? To physicists of Newton's time and after, the ether was the necessary mode of transportation for waves of light. After all, if light was a wave—as most people at the time thought—then there had to be some medium for it to wave through. Waves of wheat shimmering in the sunlight, waves of sound carrying other people's cell-phone conversations to your ears, friendly waves of hands in parting or greeting, waving flags, waving ropes, waving hair—all depend on something to carry the wave. A wave without something to wave is unthinkable.

But what could wave to carry light?

The answer was the luminiferous ether. Ah, but what queer properties this ether must have! Since the ether needs to pervade everything that light pervades, it must be able to travel everywhere. To vibrate as fast as light, Newton calculated, it must be 490 billion times more elastic than air. At the same time, planets must plow through it with no resistance, so it must be at least 100 million times less dense than air.

In short, it would have to be as hard and elastic as steel, yet completely insubstantial and undetectable.

There *was* one way to detect the ether: If Earth moved through the ether, as it must, then the motion of the planet should create an ether "wind," just as a car moving through still air creates a "wind" by its own motion.

In the 1880s, Albert Michelson and Edward Morley devised an experiment to detect this wind. Simply, they reasoned that Earth moving through the ether was a little like a person swimming through water. A swimmer traveling along with the current should go faster than someone traveling against the current. Moreover, someone swimming across the current should drift a bit during the voyage.

In the same way, the ether wind should show up in the difference in the travel time of light beams traveling parallel to, or perpendicular to, the path of Earth through the ether; one beam should also drift relative to the other.

Of course, the difference would be so small it would require an extraordinarily sensitive detector. Michelson and Morley made use of the properties of light itself. Like any other wave, light undulates up and down (to choose an arbitrary direction). The up and down

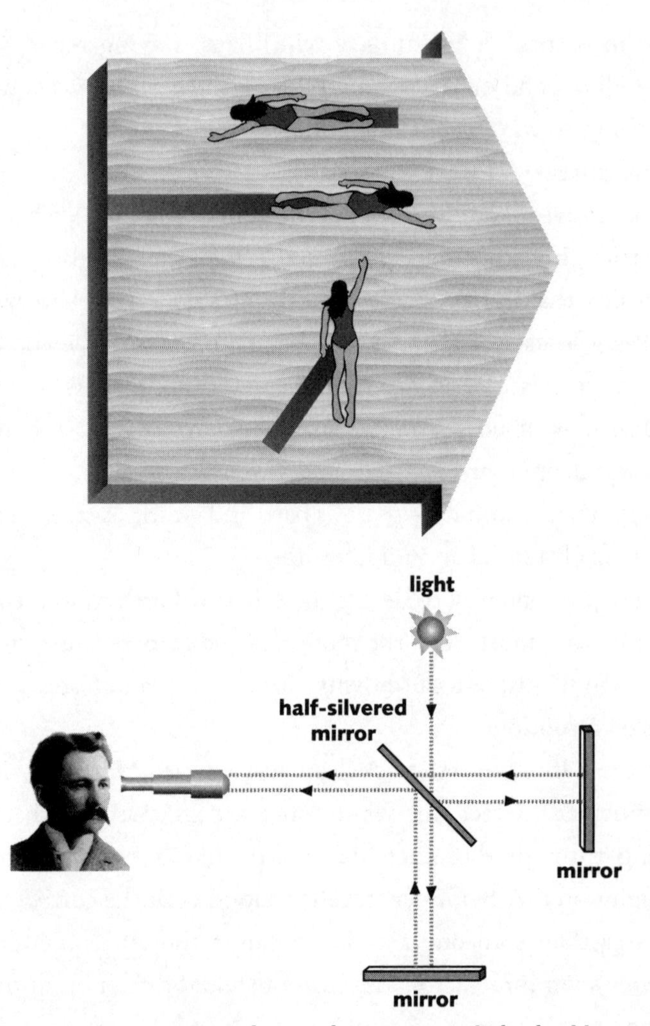

Just as a swimmer gets swept along with a current, so light should get swept along with the current created by the Earth's motion through the ether. Alas, Michelson's experiment, depicted above, failed to detect such an " ether drift."

motions of any two beams can be in sync (in phase) or out of sync (out of phase). While light waves themselves aren't visible, the patterns of in-phase and out-of-phase vibrations are readily apparent. In fact, they give rise to the colors of soap bubbles and oil slicks, among other things.

To sense differences in phase caused by an outside influence such as the ether wind, the two beams have to begin their journey exactly in step. Michelson used a half-silvered mirror to divide one beam into two identical beams. One of these beams then traveled in one direction through the supposed ether wind, while its twin traversed a perpendicular direction. Rejoined after their travels, the patterns of the beams should have been visibly out of sync.*

Alas, they were not. Michelson and Morley saw no effect. Many other experiments conducted in the late 1800s and early 1900s came up equally empty-handed. The ether did not exist.

So firmly embedded was it in the hearts of some physicists, however, that even then they were loathe to give it up—not surprisingly, Michelson in particular, had trouble letting go. "The ether was so dear to Michelson," writes physicist Marcelo Gleiser in *The Dancing Universe,* "that to the end of his life he held on to it. . . . Even as late as 1927, in his last publication, Michelson refers to the ether in nostalgic words: 'Talking in terms of the beloved ether (which is now abandoned, though I personally still cling a little to it) . . .' "

EINSTEIN'S ETHER

The introduction of a "light ether" will prove
to be superfluous, inasmuch as the view to be developed
here will not require a "space at absolute rest"
endowed with special properties. . . .

—ALBERT EINSTEIN, *On the Electrodynamics of Moving Bodies*

Albert Einstein, of course, is known as the scientist who finally swept the ether out the door and into obscurity. It's not even clear whether Einstein knew about the Michelson-Morley experiments,

---

*The inertial guidance system of a Boeing 767, I was delighted to learn on one cross-country trip, uses a modern version of Michelson's gadget to sense changes in the jet's orientation as it travels through the air.

but it didn't matter. His theory of relativity dispensed with the need for ether. Light didn't need a carrier, or even space, to travel through. An endless undulating train of electromagnetic vibrations, it pulled itself up by its bootstraps as it traveled along. What's more, space and time, in Einstein's universe, are secondary properties. The speed of light is central, solid; space and time are neither. With no absolute grid of empty space, there is no need for an ether, luminiferous or otherwise.

Einstein may have buried the ether, but ironically, he also came to praise it. In fact, in his later writings, he suggests that his four-dimensional spacetime comprises a new sort of ether, refurbished and rehabilitated. "According to the general theory of relativity," he wrote, "space without ether is unthinkable; for in such space there not only would be no propagation of light, but also . . . no basis for space-time intervals in the physical sense. But this ether may not be thought of as endowed with the qualities of ponderable media."

As this all too brief history of emptiness and ether demonstrates, the attempt to pin down the exact nature of something and nothing produces a broad spectrum of almost ghostly intermediate states which might fall into one—or often both—categories: space and ether are only the beginning. The situation gets worse—or better, depending on your point of view—as time and space go on.

Like the ether, nothing is both substantial and elusive, essential and superfluous, the most abstract and concrete of notions. Some people will always believe that Einstein's work effectively "disappeared" the ether, turning something into nothing. However, modern physics appears to be moving against the wind, in the other direction, turning nothing into a newly founded and fancied-up ether.

As Wilczek concludes: "There is a myth, repeated in many popular presentations and textbooks, that Albert Einstein swept [the ether] into the dustbin of history. The real story is more complicated and interesting. . . . Einstein first purified, and then enthroned, the ether concept. As the twentieth century has progressed, its role has only expanded. At present, renamed and thinly disguised, it dominates the accepted laws of physics."

Chapter 3

GOOD FOR NOTHING

Nothingness lies coiled in the heart
of being—like a worm.
—Jean-Paul Sartre,
Being and Nothingness

Nothing landed in the lap of humanity with a splash, and like many splashes, it created far-reaching waves. In their wake, new ideas began to germinate, blossom, and spread.

Nothing didn't start out very promising, as we've seen.

Eventually judged as too beautiful to be true, nothing initially was rejected because it was too true to be beautiful. The truths it revealed were too unsettling to fit into the prevailing scheme of things, steeped, as they were, in an abiding reverence for the perfection of God and mathematical logic.

Exactly what nothing did to become notorious is worth a short digression here precisely because—as is so often the case with problem children— the very properties that made the idea initially unpalatable became the source of much of its later power.

NOTHING BUT TROUBLE

Dividing by zero is the closest thing
there is to arithmetic blasphemy.
—WILLIAM DUNHAM, *The Mathematical Universe*

People created nothing because they needed it as a tool, and nothing, in turn, created havoc. Almost as soon as it appeared on the scene, it started spinning out paradoxes, posing unsolvable problems, and unleashing a Pandora's box of controversy, complexity, confusion.

Take the number zero. From the outset, it was an outlaw. Zero alone does not obey the rules of simple arithmetic. It refuses to behave like other numbers. Used to multiply, it erases everything; used to divide, it explodes. No wonder mathematicians sometimes think of it as The Terminator of numbers.

Of course, simply adding zero to another number doesn't change it. Same with subtraction. Zero—as far as addition and subtraction are concerned—is a complete, well, zero. It adds nothing whatever to the equation.

But multiplication—which, after all, is simply adding over and over and over again— transforms zero from Jekyll to Hyde. Multiply anything by zero, and it disappears in a puff of nothing. No matter what you start with, zero turns it into a blank slate—with no memory of its former self. "Zero destroys multiplication," says David Bayer, a mathematician from Barnard College. "Zero is very greedy. You multiply anything by zero you get zero, no matter what."

Division is even worse. Division by zero scatters whatever it touches into terminal ambiguity. It doesn't mean anything. It means everything and nothing at the same time. Division by zero is the only operation in arithmetic that is outright banned. Try to divide by zero on a simple calculator, and your calculator will scold: "Error!"

In a nutshell, here's the source of the problem: Just as subtraction is the opposite of addition, division should be the opposite of multiplication. Multiply 5 times 6 and you get 30. If you change your mind and want your 6 back, no problem. You can undo what you just did simply by dividing 30 by 5. Presto: 6. This always works because—by definition—the number you are dividing *into* is always the product of the number you're using to divide and the result. Divide 5 into 30 and the result is 6. Reverse the process and 5 times 6 is 30. It always works, backward and forward, as many times as you like.

But not with zero—because multiplying by zero always gives you zero, and only zero, back. The answer, in other words, is meaningless. Dividing by zero is like breaking an egg: You can't get back what you started with, and you always wind up with a mess.

Indeed, there's almost no mathematical operation that division by zero doesn't turn to mush. Take a real no-brainer, like zero divided by zero. A number divided by itself is always one, right? Not in zero's case. Zero divided by zero can give any answer you like. Rather than destroying information, it produces too much. After all, the number you want to "retrieve" from multiplication when you divide zero into zero is zero. And all you need to do to get zero is multiply it by any other number. So the result of dividing zero by zero can be any number in the number system, from one to infinity. Zero divided by zero can be 45, because 45 times zero equals zero. Whatever.

Dividing zero by zero gives an infinite number of answers. That doesn't mean the answer is infinity; it means you can't *say* what it is. It's completely ambiguous. It conveys about the same information as playing all the radio stations at once so that you hear nothing but static.

ENDLESS NOTHING

Like a sharp rap on the wrist from an old-time schoolteacher, an infinite answer is nature's way of telling us that we are doing something that is quite wrong.
—BRIAN GREENE, *The Elegant Universe*

Zero is Janus-faced—looking toward nothing on one side, everything on the other. And if anything, infinity causes more problems than zero. At least zero is a definite stopping point. Infinity has no edges. It leaks, unbounded. Unlike the tidy knot of zero, infinity can't be cornered. Infinite empty space can't be contained by walls or reason. One of the reasons Aristotle so hated the void was that it implied the possibility that objects would meet no resistance on their travels and could move at infinite speed. Among the Greeks, *horror infiniti* was almost as great as *horror vacui.*

Infinity, so far as it was allowed at all, belonged only to God. To think—much less say—otherwise was blasphemy. Giordano Bruno was burned at the stake in 1600 for arguing that the universe was infinite, filled with an infinity of worlds. Johannes Kepler—who finally figured out the correct elliptical shape of planetary orbits and laid the ground for Newton—was terrified of infinity. "In this immeasurable space we would be all but lost," he wrote. "It has no center and no boundaries and no specific location can be fixed in it."

Infinities plague physicists in many different contexts. In fact, when they appear in a theory, they are widely regarded as symptoms of a probably terminal disease. Infinity is nature's way of telling physicists they are on the wrong track.

Most of these infinities, one way or the other, result from dividing by zero. For example, point particles like electrons have zero dimension, zero size, zero radius. This means that all sorts of properties of the electron can add up to infinity.

Look at it this way: A beam from a flashlight fades as you move away from it. In the same way, gravity and electric charges fall off as you move away from the source. The reverse happens as you get close to the source. The force gets stronger. If you can get infinitely close, then the force can get infinitely strong.

Infinities also pop up in the hearts of black holes, where gravity is so strong that spacetime curls in on itself, in effect, shutting out the rest of the universe. These swirling pits of pure spacetime collapse down to a point of zero size. That means that anything packed inside is squeezed into infinite density.

But is there such a nonsensical thing as infinite density packed into zero size?

"I remember puzzling about that when I was a kid," says physicist Gary Horowitz of UC Santa Barbara. "I thought when I went to college I would find out the answer. And to my surprise, I'm still waiting."

There are several ways to skin an infinity. Sometimes, physicists can avoid the infinite consequences of dividing by zero through recalibrating "zero" to mean very, very, very, very small. Or they use one infinity to cancel another.

So important are tools to tame infinities, in fact, that the Nobel Prize in physics for 1999 went to two Dutch theorists who figured out how to hide the infinite answers in an otherwise promising theory. It was only the most recent in a string of Nobel Prizes for similar feats: The late Caltech physicist Richard Feynman won his Nobel for getting rid of the infinities in an earlier theory.

In some large measure, string theory owes its success to its ability to conquer infinities and the problems they cause. String theory does away with point particles. The tiny loops of string are the smallest allowable size. And by getting rid of the particles of zero dimension, string theory also gets rid of the troubling infinities.

ZERO'S GHOSTS

What are these . . . evanescent increments? . . .
They are neither finite quantities, nor quantities
infinitely small, nor yet nothing. May we not call
them the ghosts of departed quantities?
—philosopher GEORGE (later BISHOP) BERKELEY

Combining the worst of both worlds is the notion of something that's infinitely small—that is, as close to nothing as you can get, without actually being nothing. Zeno, Aristotle, and others refused to give the concept any credibility. If division into infinitely small pieces were possible, they reasoned, nothing in the universe could move. An arrow, for example, could not fly through the air, because it would have to travel through an infinite number of infinitely small points. That would take an infinite amount of time. Therefore, it must be impossible.

Indeed, the inability to deal with the infinitely small stopped people from figuring out how to deal mathematically with change for thousands of years. A chicken could be on the left side of the road or the right side. But a chicken couldn't change its position without traversing an infinity of points. In the same way, blood couldn't flow, planets couldn't orbit, tides couldn't rise. Mathematics could deal only with static geometrical objects.

Newton and Leibniz solved this problem by inventing the calculus—the mathematics of moving targets. Calculus is a concrete method for grasping the inherent slipperiness of the infinitely small; it allows you to add up an infinite number of things and get a finite answer—whether it's an infinite number of angels dancing on the head of a pin or the infinite number of points on the path from here to there.

Calculus does this by setting up zero as an unattainable goal, an elusive limit that can be snuggled up to, but never quite touched, never allowed to contaminate the calculations. Instead of zero, calculus relies on infinitely small (but not zero-sized!) slices known aptly as *infinitesimals*—arbitrarily small numbers that can do everything zero itself cannot. An infinitesimal is bigger than nothing, but smaller than anything. David Berlinski, in his enchanting *Tour of the Calculus,* compares calculating with infinitesimals to "putting butterflies on leashes."

Needless to say, not everyone thought this idea made sense. What is the difference, after all, between something of zero size and something infinitely small? Does something that's infinitely small even exist?

Bishop Berkeley, for one, didn't think so.

The infinitesimals that lie at the heart of calculus, he thought, were as insubstantial as homeopathic remedies. All that's left is the memory of what once was. That's why Berkeley famously dismissed these wanna-be zeros as "ghosts of departed quantities."

Despite the insubstantiality of infinitesimals, they are behind the ability of calculus to tame the infinite—putting those angels on a pinhead to work in the service of science, and just about everything else.

NOTHING MAKES GOOD

Lack has a prodigious propensity to meaning. He seems to attract it like a lightning rod. . . . Lack is a verb both active and passive; an object and a space at once.

—JONATHAN LETHEM, in *As She Climbed across the Table,* a novel whose hero—Lack—is a literal physical nothing.

The invention of the calculus showed just how effectively nothing's bad habits could be put to good use. Slowly, nothing shook off its unsavory past and began to contribute to society.

The first step in this turnaround, as we've seen, was getting nothing off the counting board and into the mind, where people could play with it freely, roll it around in their heads, teach it tricks, and put it to all sorts of previously unknown purposes. No longer did numbers have to be thought of as adjectives: two butterflies, three stars, four stairs. Zero was clearly nothing of anything. Other numbers took the cue, releasing themselves from the bonds of objects, free to be. Material stuff ties you down, but ideas can take you anywhere.

"[Zero's] invention liberated the human intellect from the prison bars of the counting frame," as Lancelot Hogben put it in the classic *Mathematics for the Million.*

This newfound freedom allowed mathematicians to get to know numbers out of context. They set to work creating general rules for solving problems, generic choreographies of calculation—those dances with numbers called algorithms.

Soon, the number zero and the nothing it stood for became recognized as essential tools—unreasonably effective in a wide range of contexts.

For example, nothing set origins and goalposts, pivot points and frames of reference. It gave things a clear place to start. Everything, after all, has to start somewhere: the hour, the day, the millennium, the temperature, the odometer, the ruler, the universe. Nothing provides a way for things to start from scratch. "If there was no zero," states the Zero Home Page, "cool sports cars couldn't go from 0 to 60 in xx seconds flat, now could they?"

Not all starting points are absolute, however. Absolute zero temperature, as its name implies, is usually considered a terminal condition; there's only one "true" zero degrees. Never mind that in practice this zero remains forever aloof, an unattainable tease. In theory, it's absolute and unique.*

*Absolute zero temperature, if you could get there, would be –459.67 degrees on the Fahrenheit scale, or –271 centigrade.

Not true, however, for the zero mark on your bathroom scale. Feeling a bit weighty one day? Simply slide the zero point, say, five pounds toward the negative. Presto! You're five pounds lighter. In the same way, we slide the zero in the centigrade scale way up to the freezing point of water, even though the "real" zero remains down around –271 degrees.

Sliding zeros are like odometers on cars; they can forever be adjusted. Absolute zeros are more like the big bang—a once in a lifetime thing.

Absolute zeros are more the exception than the rule. Most nothings are really the sum of equal and opposite somethings. Plus one and minus one. Or the many trillions of negative electric charges in your body, almost exactly countered by the same number of positive charges, which leaves you—most of the time—with an electric charge of zero.

And most of the time, it doesn't matter which kind of zero it is. Even if the minimum point of something is not zero, it's almost always convenient to set it to zero. Zero is a benchmark, a clear signpost that says: "You are here."

Once the origin is fixed, you can begin to roam backward and forward, away and toward ground zero, climbing up and down the x and y coordinates on a graph. Zero is the pivot point between positive and negative, yesterday and tomorrow, countdown and blastoff.

Because setting things to zero was so useful, mathematicians came up with all sorts of tools for looking for zero points that might not be otherwise obvious, for finding maxima and minima (the zero points at which values don't change). For mathematicians, scouting out zeros has become a science in and of itself.

Physics and other sciences borrow a variation on this theme. Physicists are always simplifying problems by setting complicating factors to zero: air resistance, the motion of the earth, the gravitational perturbations of planets. By counting them as zero, they sweep them under the rug.

It's difficult to overstate the utility of zeros as benchmarks, or the chaos that often ensues when they are misplaced, or missing. For want of a zero, the modern calendar became so confusing that people can't even agree on when to celebrate the millennium: According to *Scientific American* magazine, the National Institute of Standards and Technology, the Library of Congress, and the Royal Greenwich Observatory, the millennium begins with 2001. According to almost everyone else—not to mention the natural human predilection for nice round numbers—it began at the instant 1999 became history.

The reason for the continuing dispute arises from the fact that the people who came up with our modern calendar didn't have a zero. As a result, our calendar goes directly from minus one to plus one without stopping at go: 1 B.C. to A.D. 1. In effect, it goes directly from forward to reverse (or vice versa) without stopping in neutral. There is no place to turn around. Christ, like the calendar based on his birth, was born at full throttle—already a year old when he (and it) first appeared on the scene. "Babies are zero years old until their first birthday," writes Stephen Jay Gould in his aptly titled book, *Questioning the Millennium: A Rationalist's Guide to a Precisely Arbitrary Countdown.* "Modern time was already one year old at its inception."

NOTHING IN THE BACKGROUND

Imagine a billiard ball as the only inhabitant of the universe. What position does it have? The question has no meaning, for position can only be defined with respect to another position, which we call an origin, and there is nothing to define where the origin is.

—B. K. RIDLEY, *Time, Space, and Things*

To the extent that people believe in coordinate systems (treat them with respect, Steven Weinberg might say) they become real. Like

Peter Pan's Tinkerbell, they come to life when people embody them with faith.

Broadly speaking, this applies to two- or three-dimensional reference frames as well as zero-dimensional starting points or one-dimensional starting lines. An "origin" can be a two-dimensional blank page, for example, or three-dimensional empty space, or even four-dimensional space and time. The origin is the background from which all counting begins, from which everything else emerges.

Plato—who so rigorously avoided the void—still thought a great deal about the importance of this unseen background, as we have seen. Sounding a little like a chemist, Plato seemed to view the background as a kind of neutral solvent—something that allows other things to come to be without imposing too much of its personality on the result. The background can't have any qualities itself. If it did, "It would be showing its own face as well," he writes in *Timaeus.*

"This is why the thing that is to receive in itself all the [elemental] kinds must be totally devoid of any characteristics. . . . Think of people who make fragrant ointments. They expend skill and ingenuity to come up with something just like this, to have on hand to start with. The liquids that are to receive the fragrances they make as odorless as possible. Or think of people who work at impressing shapes upon soft materials. They emphatically refuse to allow any such material to already have some definite shape. Instead, they'll even it out and make it as smooth as it can be."

Newton gave the background its first solid form. As we discussed in the previous chapter, Newton's space and time provided a fixed framework for everything in the universe—a place and time to put things. He gave the universe a stage. And even if that stage turned out later to be mere scaffolding for something far more interesting, it was a major accomplishment at the time.

"There are no stationary milestones or other markers cemented into space against which to recognize rest," Banesh Hoffmann reminds us. "How, then, can we give cosmic sense to the ideas of rest and motion in a straight line, now that we no longer have a fixed earth? There is no solution. But Newton needed one, so he invented one."

Before Newton, there was no slate on which the rules of nature could be clearly drawn, patterns seen. Newton created it in his mind. As he famously declared: "Absolute space in its own nature, without relation to anything external, remains always similar and immovable."

From then on, things could move around the universe against a fixed backdrop. Position had meaning. Fixed space and time provided a grand origin from which everything else could roam, always tethered to something solid.

Today, Plato's neutral base, Newton's absolute space, Michelson's ether—all have been replaced as nature's backdrop by the afterglow of creation itself. "We now have this [new] frame of reference of the universe," said physicist Johann "Jan" Rafelski, author of *The Structured Vacuum.* "We all move with respect to the Big Bang." Called the cosmic microwave background, this preferred reference frame is the old light that still pervades space from a time when the fire of the big bang became suddenly transparent—when the universe was a mere 300,000 years young. (More on this later.)

The big bang isn't everyone's reference frame of choice. The cop who clocks your speed on the highway at 80 mph, for example, only cares how fast you're going relative to the asphalt—not the universe. For everyday purposes, most people find it convenient to set the motion of the earth, the solar system, the galaxy, to zero.

ALL YOU NEED IS NOTHING

Omnibus ex nihil decendis sufficit unum.
(One suffices to derive all out of nothing.)
—LEIBNIZ

Starting with zero, you can easily create everything. Three hundred years ago, Gottfried Leibniz became so smitten with the idea that all the numbers could be generated simply by combining zero and one that he attributed this miracle to God. As the eighteenth-century mathematician Pierre Laplace recounts it: "Leibniz saw in his binary arithmetic the image of Creation. . . . He imagined that Unity represented God, and Zero the void; that the Supreme Being drew all beings from the void, just as unity and zero express all numbers in his system of numeration."

With a little help from one, zero created the universe.

Today, people are accustomed to the idea that everything comes from ones and zeros. All of binary math used by computers is based on this system. This is not accidental. Only one and zero have a special property essential to using numbers for logical operations. That is, only one and zero don't change when multiplied by themselves. The square of zero is zero. The square of one is one.

As explained by the nineteenth-century mathematician George Boole, who invented the algebra of logic: "The symbols of Logic are subject to the special laws, $x^2 = x$. Now of the symbols of Number there are but two, viz. 0 and 1, which are subject to the same formal law."

On this basis, Boole realized that only zero and one could be the basis of his mathematics of thought.

Of course, binary computers don't really know from ones and zeros. As electronic devices, they know only from voltages. So in this sense, computers really do their magic based not so much on ones and zeros as on the even purer polarity of something and nothing, on and off. This endless on-again, off-again conversation between the simple something and pure nothing drives your car, flies your plane, operates your microwave, camera, television, clock.

(Needless to say, to my computer, this entire book is nothing but ones and zeros.)

And mathematicians didn't stop there. It wasn't enough to create everything from one and zero. How about creating everything from zero itself? Why bother with one when all you really need is zero?

Mathematicians tend to be enthralled with the idea that all the natural numbers (and therefore, everything based on numbers—which is virtually everything) can be created from nothing. As with the logic that physicists use to create the universe out of nothing, this may strike some readers as a bit of a tautological trick. Still, it's rather fun.

First, you have to convince yourself that there's such a thing as an empty set—which has a real existence, real enough, anyhow, to be counted as a true "something." The empty set could be the set of all infants with physics degrees, or moons made of green cheese, pigs that fly, or honest politicians. Whatever is in the empty set, the number of members is zero. You could call it the set of things that can't exist. Or simply a set with nothing in it: no begonias, no space, no quantity. It is a formal collection of things that are not and cannot ever be.

Either way, it's special, unique unto itself. There's only one like it. And that's the key: The number of empty sets is exactly one. So starting with the empty set (zero), and the set of empty sets (one), you can build up all the numbers from scratch. Two is the number of elements (that is, zero and one) in the two previous sets. Three is the set of zero, one, and two. And so forth. Ad infinitum.

Various mathematicians have elaborated and refined this idea over the past hundred years—most famously John von Neumann. But the point is, once you get all the positive numbers and zero, you also get all the negative numbers and anything else you want—the whole number shebang.

NOTHING ADDS UP

A man has nothing, either because he really has nothing, or because the amount of money put aside in his safe is exactly equal to the sum of his debts.

—ALBERT EINSTEIN and LEOPOLD INFELD,
The Evolution of Physics

Where there's a zero, there's almost always a positive and a negative. This seemingly simple idea opened up yet another whole career for zero—as the sum of many plus and minus parts. Zero marked the line in the proverbial sand, making positives and negatives possible. The rest is the history of the universe.

There can be a big difference between zero as a rock-bottom minimum (as in temperature) and zero as a midpoint on an infinite scale that goes both ways. Think of a bird perched on a high-voltage line. It doesn't flinch, even though the power line may be carrying a half a million volts. Sending a charge through the bird would require a *difference* in voltage between its two legs. And in this case, the difference is close to zero. Contrast this with the freezing of water. Ice doesn't freeze because of a "difference" in temperature. It always freezes at the same temperature*—no matter where we choose to set the zero on the scale.

The power of these "net" zeros can be truly astounding. For openers, we wouldn't have negative numbers without them; we wouldn't have *x* and *y* coordinates dividing positive and negative territory.

---

*At constant pressure, at least.

"Net" zero motion results from a delicate balance of forces.

It's not just about numbers. In physical terms, these opposites add up to even more impressive feats. Two waves moving in opposite directions—say, one up, one down—can add up to a flat line, no wave at all, as one exactly cancels the other. This means that two sound waves exactly out of phase can produce silence, and two light waves can produce pure dark.

In a variation on this theme, a dancer can balance in precarious postures because the pulls on all sides of her center of gravity add up to zero. This center of gravity is a point of zero dimension at which all the weight is concentrated. Nothing is there, and everything is there. The center of gravity doesn't even have to fall within the dancer's body. So long as the net force pulling on her is zero, nothing will happen. She can hover endlessly so long as she can keep her center over her toes.

Nothing doesn't have to be nothing after all. It can be an equilibrium point where things are in balance. Sometimes, this equilibrium is shaky—as when the balance of trade between two countries is built on politically charged tariffs subject to congressional whims, or when the dancer centers all her weight on a single square inch of floor.

Other equilibrium points are more stable: Two 500-pound wrestlers can push on each other with enormous (but equal) force, and they will not move an inch. The shoes in your closet don't go anywhere of their own accord, because the pressure of the earth pushing up exactly balances their weight pushing down. Bridges stand up for the same reason.

Much hides beneath the surface of seemingly innocent zeros. Like featureless eggs, zeros crack open to reveal fascinating inner structure—yolks and whites and maybe even a tiny embryo. Take a neutron, a particle with zero electric charge. It's really composed of three charged particles: one up quark with a positive electric charge of $^2/_3$, and two down quarks, each with a negative charge of $-^1/_3$.

This adding of opposite forces and motions to arrive at zero can also be turned on its head: You can get force or motion seemingly out of nothing.

Essentially, that's how a rocket gets around (or a balloon full of air after you let it go). Something rushes out the back. Something moves forward. Overall, though, nothing has changed. If you add up the change in forward motion and the change in backward motion, the result is zero.

Enormous electrical forces can be squeezed out of nothing in much the same way. Take all the electrons and protons in your body and move them one foot apart. The energy would power a good-sized city for months, perhaps years. Separate the positive and negative charges in clouds, and create lightning. Cut a magnet in half, and get two entirely new magnets, each with its own north and south poles. In effect, four poles take the place of two. Where nothing existed before, two new wellsprings of force are added to the universe.

Indeed, much of what goes on in the universe is a coming together and breaking apart of quantities that add up to zero—pluses and minuses cavorting around in a continual dance, coming and going, joining forces, then separating again: Electrons and "holes" in semiconductors, positive and negative charges, action and reaction, push and pull.

Given all this, it's not really surprising to learn that physicists have figured out a way to get everything in the universe out of nothing. There will be more about all this later in Chapter 7, "Nothing Becomes Everything." For now, suffice it to say that very little in our universe is actually nothing. Almost all the seeming nothings are actually sums of opposing somethings. Nowhere is this more evident than in the vacuum itself—what people call empty space. What seems like a silent sea of nothing at all is really an infinite number of positive and negatives, all joining together and splitting up in an endless jumble of uncertainty.

Empty space is nothing only on average. It's nothing in the sense that a billionaire with a billion dollars of debt has nothing to her name. What goes out is the same as what goes in. But that doesn't mean there's not an awful lot going in and out of her wallet in the meantime.

When things don't cancel, we know about it. Gravity exerts its huge pull on the universe, for example, only because it adds up exclusively. Gravity is a weakling among the natural forces, a trillion trillion trillion times weaker than electromagnetism. A few electrons rubbed off the surface of a balloon creates an electrical force big enough to hold the balloon on the wall in defiance of the pull of gravity from the entire earth.

But gravity is also a one-way street. Gravity *only* attracts. As a result, on very large scales, gravity can overwhelm every other force in the universe, crushing even stars into oblivion.

Still, it's sometimes the tiny deviations from perfect cancellation that have the most impressive effects. Matter and antimatter were produced in the very early universe in equal, but not exactly equal, amounts. Today that small proportion of leftover particles comprises all the matter in the universe—including us. How this tiny imbalance came to be remains a mystery. Things that add up to zero are a lot easier to explain than things that should, but don't, or almost do, but not quite.

ZERO THE CONSERVATOR

There is no theorem that says that the interesting things in the world are conserved—only the total of everything.
—RICHARD FEYNMAN, *The Character of Physical Law*

A big nothing that underlies much of the physical world is the "conservation law." A conserved quantity is one that doesn't change, no matter what. It's that great set of scales in the proverbial sky that makes sure all the pluses and minuses balance out to nothing in the end. Nothing ventured; nothing gained. At the end of the day, you can change everything, and still nothing changes.

A conservation law is nature's way of saying the more things change, the more they stay the same. The things that are conserved, not surprisingly, are the most fundamental: energy, momentum, electric charge. The universe's big zeros.

In fact, every example of a net zero discussed above contains a conservation law: rockets and balloons conserve total momentum; clouds conserve total electric charge; the universe conserves total energy. "There is a number, the total electric charge in the world, which, no matter what happens, does not change," says Feynman. "If you lose it in one place you will find it in another."

Physicists have such faith in these laws that when one of them doesn't seem to be working, they rarely blame the law. Rather, they suspect that something is going on right under their noses that they've never suspected before. The discovery of the neutrino, for example, was derived from just such a deduction.

It's more than curious how the most fundamental things in physics are those that don't ever change. In addition to conserved quantities are constants: the speed of light, for example. It's because the speed of light never changes that space and time are elastic.

Surprisingly, everything in the universe moves through the four-dimensional fabric of space and time at exactly the same speed—the speed of light! No matter what is moving, or how fast, change in motion through spacetime remains absolutely zero—no matter what.

How can this be? It works because you can divide up motion between space and time, so long as the total amount of motion doesn't change. If you're standing still, all of your motion is through time. To a close approximation, that's the situation in which we normally find ourselves. But if you're traveling very fast, some of your motion through time is diverted into your motion through space. This actually happens. It's only because our motion through time is much faster than our motion through space that we normally don't notice the inescapable trade-off. But anything traveling close to light speed through space experiences a considerable—and quite noticeable—slowdown in time.

In fact, particles that travel at exactly light speed—for example, the light particle, or photon—use up all of their motion in speeding through space. A photon, therefore, does not travel through time. "Thus light does not get old," explains Greene. "A photon that emerged from the big bang is the same age today as it was then. There is no passage of time at light speed."

Nothing changes. Nothing happens. Nothing matters. These nothings are the basis of the fundamental symmetry of nature—the smooth surface of perfection that contains the potential for everything. Nature's sweet nothings harbor the seeds of all there is.

Chapter 4

NOTHING TAKES CENTER STAGE

We now know that there is, in principle, no permanence in substance; it is mere bottled energy, and possesses no more inherent permanence than bottled beer.
—SIR JAMES JEANS, *Physics and Philosophy*

IT ALL SOUNDED so logical. Nothing can come from nothing. As Lucretius so eloquently argued: "If things were made out of nothing, any species could spring from any source and nothing would require seed. Men could arise from the sea and scaly fish from the earth, and birds could be hatched out of the sky. Cattle and other farm animals and every kind of wild beast, multiplying indiscriminately, would occupy cultivated and waste lands alike.... Tiny tots would turn suddenly into young men and trees would shoot up spontaneously out of earth."

Lucretius and everyone else concluded the obvious: "It must therefore be admitted that nothing can be made out of nothing."

Now we know better. Or, at least, we know different. Or, at rock bottom, we have gotten more sophisticated in how we discriminate "something" from "nothing."

Lucretius and his intellectual successors divided the universe into two distinct categories: things that are and things that are not. Like butterflies and chairs, it was clear that something and nothing were two different species; they weren't related, and certainly, they couldn't interbreed. The only possible progeny to be expected from nothing was more nothing; from something, more of the same.

This seemingly inarguable assumption turned out to be unwarranted. Like the birds and fish of an Escher woodcut, something and nothing can't be teased out, one from the other, to stand as separate entities unto themselves. The universe is just one big happy tapestry of tangled relationships that can never be unraveled. There is no chair here, butterfly there; particle here, void there; time here, gravity there. There is only the picture that emerges from all pulling together, a great mosaic that seems unrecognizable close up, but comes into focus as we stand back and observe from a more distant, and broader, perspective.

"Whatever the things of the Universe may be, they are certainly not like the utopian billiard ball, isolated from everything else by the glass case in which it sits," writes physicist B. K. Ridley in *Time, Space, and Things.* The glass, the billiard ball, the case, and the air inside are all of a piece.

Getting something from nothing is possible, it turns out, simply because nothing isn't at all what we thought it should be. If people are surprised at what nothing can do, it is because they didn't know how rich nothing was in the first place.

NOTHING DOING

Although atoms are way more than 99.99 percent empty space, I have a real problem in walking through a wall.

—Leon Lederman, *The God Particle*

There is a simple progression here, from things that make sense, to things that do not, but that seems to be the way it is in physics.

Take a wall, any wall. What is it made of? Mostly, empty space.

Every atom is at heart a tiny nucleus enclosed in a vast shell of pure void, bounded by ephemeral electrons. Imagine the largest room you've ever been inside—a stadium or a cathedral. Put a pea in the middle. That's the nucleus. The electrons would be buzzing around the outer walls. A large bunch of virtually empty atoms is all there is, in effect, to a wall.

Your body is the same: tiny clusters of nuclear particles, moving electrons, lots of empty space. You'd think, when two objects made out of mostly empty space collide, they would slide right through each other, like waves of sound or galaxies of stars. But instead, walls and people stop each other cold. Something in that empty space of the wall is putting up resistance to invasion by the empty space of your body.

What could it be? The short answer is: Space isn't empty; what we think of as empty space is permeated by powerful influences. Indeed, matter itself is only the energetic geometry of forces in empty space. But this short answer has a long and curious history.

The idea that matter is the geometry of space goes back at least (again) to Plato. Plato thought the elements were essentially geometric forms: water an icosahedron; air an octahedron; fire a pyramid; earth a cube. The cubical form of earth rendered it sturdy, immovable—which explained why our aptly named planet sat at the center of the universe. The sharp points of pyramids made fire hurt.

Descartes and others also played with a version of this idea. A nonbeliever as far as empty space was concerned, Descartes argued that all space was filled with a very rarefied form of matter—the same stuff as palpable matter, but extremely dilute. In the occasional puddles where this dilute stuff condensed, it became "matter"; the form of the matter depended on what shape it took. For Descartes, in other words, there was no distinguishing between matter and empty space: there was just one substance, molded into various forms. Nothing became something by sheer powers of concentration.

None of this became what you might call real physics, however, until the great nineteenth-century experimentalist Michael Faraday introduced what was to become one of the most radical ideas in science: Particles of matter were in themselves rather irrelevant; they were only the spigots through which various forces flowed. A solid object, if you like, was something like a fountain composed of intersecting cascades of water, all flowing from tiny pointlike orifices. The "real" stuff of matter was the flowing water, or forces; the particles were only the source.

Faraday, in effect, shifted the burden of reality from the particle to the things that emanate from it. The space between the particles became primary. "Force, not substance, is the true being of the world, and it, not the ether, reaches from one end of the universe to the other," explains Amherst physicist Arthur Zajonc. "Points of matter—atoms—were only the starlike intersections of myriad raying lines of force that spread out from these centers to weave their way through the universe."

These raying lines of force were known as fields—and today, fields are just about all there is to fundamental physics.

WRINKLES IN SPACE

For us, who took in Faraday's ideas so to speak
with our mother's milk, it is hard to appreciate
their greatness and audacity.

—ALBERT EINSTEIN

But if fields are everything, what are they exactly?

The most familiar image of a field is the orderly arrangement of iron filings as they line up in geometric curves bridging the north and south poles of a magnet. Faraday wondered: "How was each single iron filing among a lot scattered on a piece of paper to know of the single electric particles running round in a nearby [electromagnet]?" How could they communicate their presence across nothing at all, without so much as a tin-can telephone to keep in touch?

His answer was that the electric and magnetic forces create stresses within space itself. These stress lines are paths that iron filings can follow. They are an intersecting complex of superhighways that tell particles—and everything composed of particles—how to get from here to there.

Think of empty space as a flat piece of paper. If you sprinkle, say, sand on the paper, it will fall every which way. But if you crease the paper, you can make it into a simple funnel. Now the paper will direct the sand to flow in only one direction. That's essentially how a field operates, as a series of well-placed wrinkles that permeate empty space.*

According to this new understanding of the universe, objects under the influence of forces are not really pushed or pulled around; instead, the forces are creases in space that tell the objects how to move. Iron filings fall into place in precise patterns around the poles of a magnet not because they are attracted or repelled, but because they are following the path of the field—just as Dorothy followed the yellow-brick road to Oz.

The fields are more flexible, more active, than creases. Fields can wiggle around—especially if something comes along and gives them a nudge. "You can think of a field as a kind of jelly," says Frank Wilczek. "If you tweak it at one end, a wiggle moves through it."

If an electron wiggles in the sun, it tweaks the electromagnetic field, and eight minutes later the ripple in the field arrives on Earth to tickle an electron in your eye, allowing you to see light. Light is nothing but a tweak in the electromagnetic jelly (field).

---

*These wrinkles pervading space are not the same as wrinkles of space itself, which we will meet in the next chapter.

"Of course," says Wilczek, "these are very peculiar jellies. They can interpenetrate each other. Also, they look the same whether you're moving through them or not; that's certainly not true of ordinary jelly." (It's the fact that you can't tell whether or not you're moving in these jellies that allows them to be part of the vacuum, or nothing—more about which later.)

Jelly or yellow-brick road, the upshot was that Faraday took empty space and folded it with an interlocking network of permanent-press creases that connect everything in the universe to everything else—including matter and force.

The power of this idea surprised everyone. It seemed absurd to think that no clear line in the sand separated the universe of chairs and butterflies from the universe of gravity and magnetism. "Even Faraday's most sympathetic reviewers thought he had, for once, gone too far," writes Zajonc. "To make something as insubstantial as lines of force the ontological basis of the world seemed preposterous."

Fields as real, substantive, things seemed especially ridiculous because at first the term emerged, more or less, as a manner of speaking—a way of thinking about how iron filings could communicate with the poles of a magnet through empty space. Then the metaphor came to life; the abstract idea took on uncanny concreteness. It was almost as if you had said, "Love is like a rose," only to find out that love really *was* a rose. As Robert March puts it in *Physics for Poets,* "[T]he field concept that began so modestly as a substitute for action at a distance materializes as matter itself!"

The concept of field fudged forever the difference between something and nothing. It was a huge revolution in thought that remains completely unknown to most laypeople. And yet, it leaves physicists—to whom fields are bread and butter (and for whom bread and butter are nothing but fields)—in a state of semipermanent awe. Einstein called the "change in the conception of reality" from particles and empty space to fields "the most profound and fruitful one that has come to physics since Newton."

Or as Dartmouth physicist Marcelo Gleiser puts it, even more succinctly: "Physics has never been the same since."

MATTER AS METAPHOR

What impresses our senses as matter is really a great concentration of energy into a comparatively small space.
—ALBERT EINSTEIN and LEOPOLD INFELD, *The Evolution of Physics*

Matter, in this view, is simply a place where some of the field happens to be concentrated. Matter condenses out of field like water droplets condense out of water vapor in a steamy bathroom. Particles of matter are concentrations of field that travel through the field like ripples in a rope or a wave in water. The essential "stuff"—that is to say, the rope or the water—does not travel from place to place. Only the kink travels. Just as rumors can spread through a crowd of stationary people, concentrations can spread through stationary fields. Particles are more like rumors than people.

"*Seemingly solid matter is nothing more than a manifestation of fields that do not 'occupy' space at all,*" exults March (emphasis his). The field is not so much something *in* space, as *of* space.

This view of matter explains, among other things, why every electron in the universe is exactly the same as every other electron, every top quark the same as every other top quark. A particle doesn't really exist in its own right. It is only a particular manifestation of a field. And globally speaking, the field is everywhere the same.

And yet, a field can take on a life of its own, even when no longer connected to whatever object created it. Turn off the water, and pretty soon the vapor disappears. But fields can continue to propagate endlessly, even after the "spigot" that produced a force has been shut off.

An electron wriggling in a distant star sets up a traveling kink in the electromagnetic field. The kink takes off, leaving the star far behind. Millions of years later, the kink reaches Earth; the star, by that time, could have collapsed into a burned-out cinder, incapable of producing light. Yet the light shines on. In fact, kinks in the electromagnetic field still linger from the era just 300,000 years after the big bang—some 13 billion years ago. The patterns of the kinks have the potential to tell astronomers much about the detailed structure of the newborn universe in the days (before there were days) when it was still wet behind the ears.

The fields, in short, are substantive things—the stuff of stuff. "The field, although nearly as ethereal as the ether itself, can be said to have physical reality," says John Wheeler. "It occupies space. It contains energy. . . . We must then be content to define the vacuum of everyday discourse as a region free of matter, but not free of field."

The connection between matter and field takes on an even more concrete form when Einstein's notorious $E = mc^2$ is factored into the equation. The E is for energy; m for mass; c for the speed of light. In essence, the equation is a simple sentence that says: energy and matter are the same stuff—except that it takes an enormous amount of energy to equal a tiny bit of matter.

Thus Einstein concludes that any distinction between matter and field is purely artificial. "Matter is where the concentration of energy is great, field is where the concentration of energy is small," he writes. "But if this is the case, then the difference between matter and field is a quantitative rather than a qualitative one. There is no sense in regarding matter and field as two qualities quite different from each other."

Field becomes a bridge between matter and empty space, something and nothing, an intermediate state with the potential to take either form. Heat a piece of iron, Einstein suggests, and it weighs more than it did when it was cool. Wind up the spring on a clock, says MIT physicist Philip Morrison, and the clock gains weight.

These tiny increases are far beyond the sensitivity of any known measurement. But in many cases the seeming alchemy that turns matter into energy and vice versa is palpable, even familiar. The Sun sheds several ocean liners' worth of mass each day to keep its nuclear furnaces burning; particles pushed to nearly light speed in accelerators gain tens of thousands of times their normal weight.

NOTHING BUT FIELDS

Now we believe that the entire universe
is nothing but a field theory.
—physicist SIDNEY COLEMAN, Harvard University

Each variety of matter in the universe—and each variety of force—is a kink in its own special field, its own flavor of jelly. We swim among these fields every day, bumping now and then into kinks (like particles) that keep us from walking through walls. These various fields weave together in complex ways, sometimes interfering with each other to produce, for example, a force (one field) on a particle (another field). Their interaction accounts for everything that happens in the universe.

(In string theory—discussed in more detail later—every force and particle in the universe may turn out to be a vibration of the *same* meta field.)

If nothing tweaks the fields, they don't go away. They can't go away, because they're part of the structure of the vacuum itself. The fields in their quietest possible state *are* the vacuum. Fields are the permanent backdrop to everything that happens in the universe, waiting in the wings, so to speak, until something comes along to turn the spigot of forces on. The spigots, or creases, are there whether or not there happens to be any sand or water or force or matter to flow through them.

The fields are as close to nothing as anything ever gets. The quiet state of the field is a perfect vacuum.

Imagine you have a mattress, little springs all attached to one another. You can jump on the mattress and make it wiggle around. But sooner or later the wiggles die down. "It settles and stops," says physicist David Gross, director of the Institute for Theoretical Physics at UC Santa Barbara. "It's still pretty good as a vacuum because it does nothing. It just stays there. All the excess energy of the higher modes—the more vibrant oscillations—settle down.... When a particle physicist talks about a vacuum, he means a ground state of this . . . system."

Of course, there is a difference between matter and force field, at least as most people are accustomed to thinking about them. Forces push and pull things around; they attract, repel, bind together, break apart. Forces act on things; matter is acted upon. You cannot walk through a wall, but you can easily (and do routinely) walk through electromagnetic fields and gravitational ones.

That's because kinks in force fields behave differently from kinks in matter fields in one crucial way: Kinks in force fields can pile up on each other; they can squash together like feathers in a goose-down pillow. So you can walk through a force field the same way as you would walk through a wall of feathers, by pushing some aside as you go. Individual kinks in matter fields, however, do not move aside for one another. Standoffish by nature, they hold one another at arm's length, presenting a solid obstacle to intruders. The kinks in the matter fields that make up your body can't make way for the kinks that make up the wall.

But like so many things in the natural world that seemed to be quite distinct, matter and force—like energy and matter—appear to be kindred spirits underneath.

"The distinction between matter and forces is ancient, mythological," says Fermilab physicist Chris Quigg. "Air and water and love and strife. That separation has been useful until now. But things are telling us that it is not fundamental in the end. I think this wall between force carriers and fundamental constituents is going to fall sometime soon."

The evidence comes from many different fronts: certain force carriers called gluons can clump together to form hypothetical matter particles called glueballs. Many of the most promising theories—promising in the sense that they solve otherwise insoluble puzzles—tell physicists that force and matter are closely related. Like children dressed up for Halloween, they only appear different because it happens to be a particular day (or epoch, as the case may be) in the life of the universe. Strip off the masks, and lo and behold, they are the same.

In the end, calling one kind of kink a force carrier and another kind a particle of matter does not get us any closer to the essential nature of what is, and what is not, something and nothing. The only thing that differentiates something and nothing, in the physicists' parlance, is energy. If the field has no energy, it is truly nothing.

Alas, as we shall soon see, that scenario is not—even theoretically—possible.

NOTHING NEVER SLEEPS

Nature does not allow its constituents to be cornered.
—Brian Greene

Faraday created fields and made them real. Quantum mechanics made them magic.

Quantum mechanics is the term for the new understanding of the atom that emerged over a period of years during the early part of the twentieth century. It consists of several core, tightly interrelated ideas.

The first is simple to state, if somewhat counterintuitive: Nothing in nature happens in a continuous smooth flow. Everything comes in lumps. When you look at the natural world with fine enough magnification, all becomes grainy: energy, motion, time, mass, whatever. What appears to be continuous from a distance is a mosaic made of discrete bits.

Everything includes, of course, fields. So fields, too, are chopped up into the smallest possible currency—the quantum fields are the coins, so to speak, of the quantum realm.

The second idea is a little more fuzzy, its interpretation a lot more controversial. In short, quantum mechanics revealed that the universe is inherently uncertain. If you try to pin down a particle to measure its properties, it slips out of your grasp. The very act of holding it down to measure it destroys the properties you set out to measure—just as surely as holding a snowflake in your hand melts the ice crystal before you can study its geometry.

Of course, you can always put a snowflake on a very cold surface and snap a photograph without disturbing the delicate order of its crystal arrangement. Not so, however, with a quantum particle. Quantum order is so sensitive that any attempt to measure it precisely destroys it. If you try to measure a quantum snowflake precisely, it will always melt.

This means, for one thing, that no quantum property can ever be zero, because zero is a very precise number.* And that has enormous implications: Energy can never be zero; motion can never completely cease. Everything in the quantum world—and that includes everything there is, including nothing—lives in a state of continual agitation, squirming around like a restless child in school.

---

*Of course, it couldn't be precisely 4.5789287, either.

The harder you try to pin things down, the worse the situation gets. Say you want to trap an electron in order to measure its exact position. You put it in a box. You make the walls of the box smaller and smaller in order to pin down its exact position. In response, the electron moves about increasingly fast, making your goal increasingly elusive. The harder you push, the more surely it slips away

But what, you might well ask, if you never try to measure anything? Couldn't you hope to find—at least theoretically—some true quiet corner in the cosmos then?

There are many answers to this question, but for our purposes, two will suffice.

One answer is that things that can't be measured have no reality, and certainly no place in physics, so they can't be considered "real." If it isn't measured, it doesn't exist.

"In quantum mechanics, you can't ask a question without talking about how you measure it," says Gross. "You have to look at it. If you don't look at it, nothing happens. So we describe this as the ground state. It just stays there. It's a vacuum. But if you observe it, the ground state doesn't remain the ground state. . . . That's why it makes no sense to think of the vacuum as nothing there. Because you have to measure it."

The second way to answer the question requires a further leap of faith: When not being measured, quantum properties take on all possible values at the same time.

Imagine a spinning coin. Is it heads or tails? It's neither, and it's both. Or think of a rapidly spinning fan blade. Where, exactly, is the blade? In the quantum world, it's everywhere at once—at least until you try to measure it. Once you stop the blade, you can say precisely where it is. But in the process, you have effectively "destroyed" its velocity.

Quantum mechanics is full of these trade-offs. You can know position perfectly if you are willing to sacrifice knowledge about velocity, and vice versa. You can know energy perfectly if you are willing to take an infinite period of time to measure it. (Since you can't take an infinite period of time, of course, you can't know energy with perfect precision—another way of understanding why energy can never be precisely zero.) Conversely, you can pin down an arbitrarily short sliver of time but only at the price of spending an uncertain amount of energy.

But even when completely undisturbed, quantum reality seems to be inherently fuzzy. In Wheeler's description, an everyday macroscopic particle would be something like a pea that a young girl might load into her pea shooter to attack her brother. If she tried that trick with a quantum pea, however, it would drift toward her brother's head in the form of a huge pea pillow with its possible range of positions expanding like a cloud into a shifting haze of probabilities.

In a sense, quantum mechanics eliminated the very idea of zero from the physical world. Zero is too precise, too settled, for the fuzzy world of fundamental physics. In the quantum realm, even nothing never sleeps. Nothing is always up to something. Even when there is absolutely nothing going on, and nothing there to do it.

HOLES IN NOTHING

"Look here, Paul," the dolphin was saying, "you contend that we are not in a vacuum but in a material medium formed by particles with negative mass. As far as I am concerned, water is not different at all from empty space."

—Dolphin to Paul Dirac, in GEORGE GAMOW'S *Mr. Tompkins Explores the Atom*

When physicist Paul Dirac married quantum mechanics to Einstein's $E = mc^2$ and all that in the late 1920s, nothing got even weirder. Dirac's equations produced two sets of solutions for the energy of particles like the electron: one positive, one negative.

What could this possibly mean? It seemed to imply the existence of a "negative energy" electron. But such a species of particle had never been imagined, much less seen.

Remarkably enough, just a few years after the strange particle popped up in Dirac's equations, Caltech physicist Carl Anderson found an unusual track in a cloud chamber he was using to watch the trajectories of cosmic rays streaming from space. The particle track was exactly like that of an electron—except it curved backward under the influence of a magnetic field.

To Dirac, however, this "antielectron" was actually a very palpable hole in nothing. Dirac imagined that empty space was chock-full of negative energy electrons. We can't see this ocean of negative energy particles for the same reason that the dolphin doesn't "see" water; it's everywhere; it's featureless; it has no breaks or boundaries to make it visible.

This underworld of unseen particles had to be there, Dirac thought, to prevent ordinary electrons from giving off all their energy and spiraling down into negative energy states. With the ocean of empty space already packed full, there would be no place for ordinary electrons to fall.

But what if one of these negative energy electrons somehow acquired enough energy to jump right out of this sea of nothingness? he wondered. It would become a positive energy electron—in other words, one of the normal everyday electrons that make glue sticky and keep us from walking through walls.

Left behind in the sea of nothingness would be a hole—an exact complement, or mirror image, of the now-liberated electron. That hole was Anderson's antielectron, or positron—a particle with positive energy, but opposite to the electron in every other respect.

Dirac didn't have all the details right; now that physicists understand antimatter better, it's no longer viewed as holes in nothing.

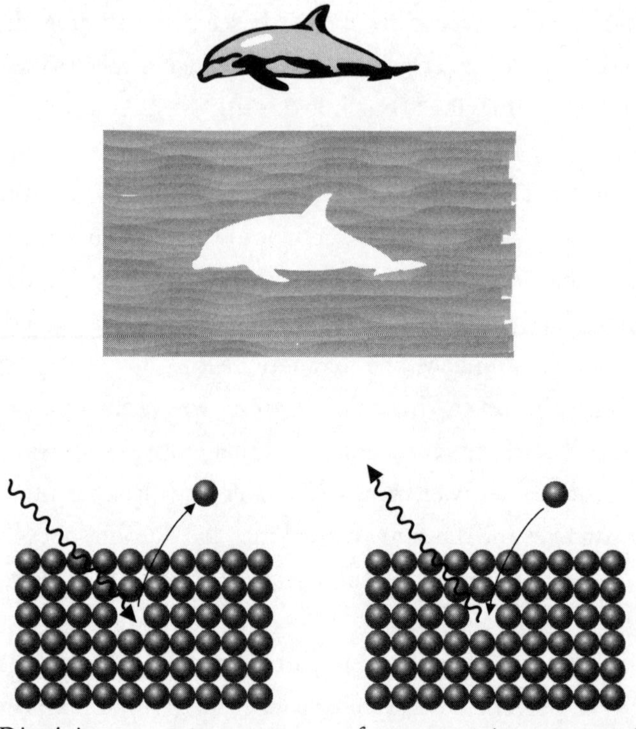

In Dirac's imagery, energy enters a sea of unseen negative energy particles ("nothing"), producing one particle and one "hole," or antiparticle (left). *When the particle jumps back into the "sea," both particle and hole disappear, radiating energy* (right).

A positron is a real particle in its own right. Still, as Wheeler points out, the modern antiparticle theory and Dirac's hole theory "differ only in their imagery, not their mathematics."

Either way, the consequences are simply staggering: Pump enough energy into the vacuum and you can create two particles where before there were none: one normal particle and one "hole," or antiparticle; or one electron and one positron. If the electron jumps back into the "sea," it quickly fills up the hole. Nothingness goes back to being nothing. Both electron and hole disappear. Both particle and antiparticle annihilate—giving back, in the process, the energy it took to create them in the first place.

LIVING ON BORROWED ENERGY

Created and annihilated, created and annihilated—what a waste of time.
—RICHARD FEYNMAN

Start with the notion that nothing can ever be precisely zero, so the vacuum is always pumped with energy. Add the idea that fluctuating fields are quantized into particle-like lumps. Factor in the knowledge that energy is the same as matter. Remember that the appearance of a particle from pure energy out of nothing must be accompanied by the appearance of a hole, or antiparticle.

What you get is a continual popping out of particles and antiparticles from nothing—creation ex nihilo all over the universe, all the time. It's a picture that leaves even hardened physicists overflowing with poetic imagery, frothing at the pen about frenzied dances of particles, roiling seas, teeming cauldrons, and seething volcanoes.

Still, strict rules govern this seeming chaos. For example, time and energy, as mentioned, are closely related. The longer you watch the vacuum, the more accurately you can measure its energy, and the closer that energy could conceivably get to zero. Conversely, the shorter the time, the less accurate the measurement. This implies that even a hugely energetic pair of particles could sneak out of the vacuum so long as they snuck back in again before you noticed—like a child on Rollerblades who could steal the cookies then put them back in less than the blink of an eye. (Although one has to wonder, as Feynman did, what would be the point?)

The cookies, like the energy, are merely borrowed. The faster they're returned, the more can be borrowed before any serious rules are violated. In the vacuum, this borrowed energy briefly comes to life as particle/antiparticle pairs ($E = mc^2$). The more energy, the more massive the particles that can be created. Particles that require less energy to create (like electrons) pop out more often. The ones that require more energy (like heavy quarks) pop out more rarely. "It is theoretically possible that a macro object such as an apple might be created for an instant," says Martin Gardner, "but the probability of this is far too low to allow it." Some people think the entire universe popped out of the vacuum, just like an apple, or a rabbit out of a hat.

But every ounce of energy stolen from the vacuum must be returned. Everything in the vacuum lives on borrowed energy. Since these particles don't hang around for long, physicists call them *virtual* particles. They come and go, like dreams. In a sense, they are the physicists' answer to Berkeley's "ghosts of departed quantities." Except virtual particles are ghosts of particles as yet unborn.

There is a way to breathe life into these ghostly particles. With enough energy, they can become real, just as the wooden puppet Pinocchio became a real boy. With enough energy, just about anything could emerge from the vacuum.

The process works both ways. Anything can disappear into the vacuum, as well. Lucretius was wrong again. "Nature revolves everything into its component atoms and never reduces anything to nothing," he wrote. But the road from nothing to something goes forward *and* backward. Real particles can vanish into nothing if they meet their mirror image in the form of an antiparticle. Think of the antiparticle as one of Dirac's holes. You start with a more or less flat surface, the gently simmering vacuum. You send in some energy and liberate a particle. You leave a hole. When the particle goes back into the hole, both hole and particle disappear, and you're back to a flat surface. The energy you put in comes out again in the form of light. It's the same process that creates and destroys virtual particles, but with far longer-lasting effects.

IMPRISONED IN NOTHING

We really got more than we opted for
from this modern ether.
—JOHANN RAFELSKI

It's clear that the new, improved vacuum does everything the discredited ether did—and a whole lot more. But this doesn't mean that physicists can't make up their minds about the nature of nothing—or the nature of anything, for that matter. One of the most common misconceptions about science (frequently fostered, I'm sorry to say, by science writers) is that scientific truth can't be trusted because it is continually being revised. *Au contraire.* It can be trusted precisely *because* it is continually being revised.

As Einstein explained it, the process of discovering new truths about nature is less like tearing down an old building to construct a new one than like climbing a mountain to get a better view. You look back from your new higher perspective, and you see things more clearly. You realize that the dense forest you climbed through is just a tiny patch of trees in a huge meadow, or you see that the "face" on Mars is a serendipitous configuration of valleys and hills. Or you see that empty space is not a Styrofoam-like ether, but something else a whole lot more interesting.

As physicists climb higher (making a good number of wrong turns and detours in the process), they see a little more of the continually emerging picture. And as they delve ever more deeply into nothing, they climb a spiral staircase that sometimes circles strangely back to ideas seemingly abandoned long ago.

This should be expected, since science is more like a running argument than a firm set of rules. But the new ideas return with new—sometimes radically new—twists.

One of the most curious qualities of the new ether is that it acts like glue; in fact, without it, the protons and neutrons that make up the very core of every atom would not hold together. Protons and neutrons are in turn made of quarks. But quarks stick together in a most peculiar fashion: the farther away they get from each other, the *stronger* their attraction becomes. This is precisely the opposite from the way, say, two magnets attract each other, or two stars. Gravity and electromagnetism get weaker as the source gets farther away. But for some odd reason, the force between quarks gets stronger.

The result is that quarks are permanently imprisoned. Try to escape, and the glue holding them together only gets stronger—infinitely strong, in fact—as the quarks get farther apart. A single quark has never been seen, a situation that is likely to continue unless physicists succeed with some radical proposals to set them free (see Free the Quarks, below).

At least one explanation for this puzzling entrapment of quarks has been found lurking—where else?—in the depths of the vacuum. Among the multitude of particles that continually froth from this turbulent sea are particles called gluons. Gluons are so named because they hold quarks together. But gluons, like quarks, have peculiar properties. They stick to one another, as well as gluing quarks together. Like water molecules sticking together to form a skin that traps air inside a bubble, physicists believe gluons in the vacuum make bubbles that trap quarks inside.

A proton or neutron "is like a bubble containing the quarks which abhor the vacuum," says Jan Rafelski. "The quarks try to get out of their bubbles all the time, but they are reflected back by the surrounding vacuum." If you looked closely enough at an atom, Rafelski says, you would see something like Swiss cheese: the cheese is the vacuum, and the protons and neutrons are the holes inside.

This "gluon ether" holds the quarks inside for the simple reason that the quarks have nowhere to go; it fills all space. Like people trapped in a crowd at Macy's on the day after Thanksgiving, quarks are pushed into close quarters whether they like it or not.

In fact, nearly all of the mass of protons and neutrons—and everything made of them—comes from this energy of nothing, according to some physicists. The ordinary quarks that make up ordinary matter weigh practically nothing at all. Gluons have no mass. But the energy of the quark and gluon fields that hold the quarks together inside protons and neutrons makes up most of the mass that "makes us weigh," as Frank Wilczek puts it in *Physics Today.* "[The field] thus provides, quite literally, mass without mass. . . . If your friend puts on a few pounds and yet complains, 'But I never eat anything heavy,' modern physics sanctions you to give her (or him) the benefit of the doubt."

The surprising power of the gluon pairs naturally makes one wonder what else is going on in all that nothingness: The delicious truth is, physicists don't know. Remember that the higher the energy pumped into the vacuum, the more massive the particles that can appear out of nothing. In the very early universe, the enormously concentrated energy would have been ample enough to produce all sorts of unknown particles—some of them hugely heavy.

The vacuum is infinitely creative. All it needs is the energy to do what it's naturally inclined to do: proliferate wildly in every conceivable form. In this sense, nothing is really the potential to do things, an idea we will revisit later.

DRESSED IN NOTHING

The "naked" electron is an imaginary object cut off from the influences of the field, whereas a "dressed" electron carries the imprint of the universe.

—LEON LEDERMAN, *The God Particle*

As pretty a picture as all this paints, the end result adds up to absolutely nothing. Every particle is created along with its opposite number—its antiparticle. The fields fluctuate this way and that, but on average, their net energy is zero. Each bit of activity is balanced by another bit of activity, which pulls in the opposite direction. Altogether, nothing.

Of course, you could say the same about lightning. At the end of the day, the negative and positive charges cancel out. But during that fleeting off-balance instability, it glows with such a lovely light!

In the same way, these nothing particles have very real effects—more subtle than a lightning strike to be sure, but impressive just the same. Indeed, the very real effects of the virtual goings-on in the vacuum have been known to physicists since the 1920s and have been proved experimentally to almost a dozen decimal places. It's a curious thought: clouds of ghostly particles, appearing and disappearing into nothingness, too fleet to be real, but real enough to be measured. "The swarm of virtual particles that make up the retinue of the electron, although invisible, lets us know they are there," writes Wheeler, in his autobiography, *Geons, Black Holes, and Quantum Foam.*

The measurements explain why physicists put such great faith in quantum theory, even though they don't fully understand it: It works too well. It's the most exact theory ever discovered (or invented) by human beings. It also explains why laypeople should resist the temptation to dismiss these unfamiliar notions as nothing more than some physicist's bad dream.

How does the vacuum make itself seen? Well, for one thing, it doesn't let elementary particles like electrons run around naked. The vacuum makes electrons as modest as women shrouded in veils; they never appear without chaperons in the form of virtual particle pairs. This thick mist of virtual particles obscures some of the electron's true or "naked" charge from view. The naked charge is actually much larger than the shielded one, and these differences can be measured.

"An electron sitting quiescent in empty space is not quiescent at all," Wheeler describes the situation. "As we zoom in on it with hypothetical microscopes of higher and higher power, we see a more and more lively neighborhood around the electron. Other electrons and positrons are coming into existence and vanishing. Photons are being born and are dying. Heavier particles join the dance of incessant creation and annihilation. And the closer we get, the more violent the activity becomes. The 'isolated' electron is the nub of a seething volcano."

The activity of this seething volcano shows up in several ways. For example, it tugs on electrons orbiting atoms, distorting the light they emit. The distorted orbits show up as slight shifts in the atom's spectrum.

"It is as if Earth, in its trip around the Sun, were to feel not only the force of attraction of the distant Sun, but also millions of tiny forces arising from tiny lumps of clay that get in its path as it cruises through space," Wheeler explains.

In a sense, every atom, every electron, carries the weight of the entire universe of empty space with it.

There are other kinds of evidence as well: The pressure of the vacuum will push two thin metal plates together. Vacuum fluctuations are waves, and as such, come in various wavelengths. If the plates are narrowly spaced, only the smallest waves fit inside the sandwich of plates. The larger ones are excluded just as surely as Alice was when she was too big to get through the tiny door to enter the beautiful garden. Since you need to be the right size to gain admission, more vacuum activity takes place outside the plates than inside their tight boundaries. As the outside pushes in, the plates get closer together.

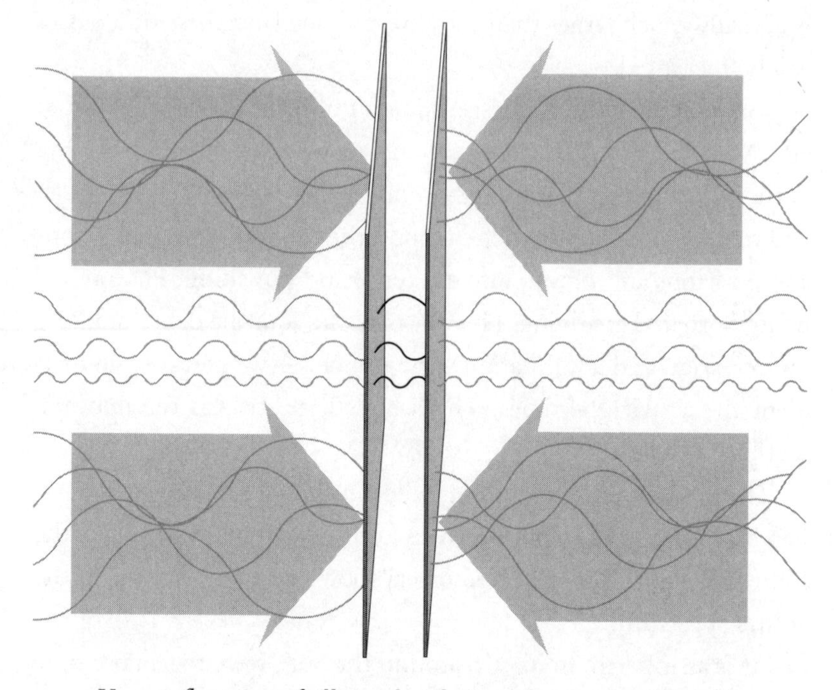

Vacuum fluctations of all wavelengths pervade space, but only a few can fit precisely within two thin metal plates. The excess vacuum energy on the outside of the plates presses inward.

FREEZE-DRIED NOTHING

Cosmologists believe that the vacuum structure of our Universe changed many times since the Big Bang.

—Johann Rafelski

Now that we know that nothing is real, let's see what it can do. It doesn't just sit there, after all, happily wiggling its fields. It grows; it changes; it even evolves. This isn't quite as difficult to understand as it is seems because it's all based on some very simple and familiar ideas: The vacuum changes its properties in much the same way as water changes when it freezes into ice.

You might as well know up front that most physicists believe we are currently living in a frozen vacuum that was once—in the very early universe—a very different "melted" vacuum. Don't try to grasp all of this at once. As Sidney Coleman likes to say, the vacuum is very complicated, rather like a jet plane. "And if I try to explain the whole jet plane at once, I'll go crazy." So one step at a time.

This frozen vacuum doesn't feel particularly cold or solid or hard to move around in because it is, after all, our vacuum—the only kind our world has ever known. It's also the only kind in which our world could conceivably exist. The structure of the vacuum (its permanent-press wrinkles) determines what can happen in our universe: which particles can appear (given enough energy), which forces can push them around, maybe even the nature of time and space.

How can frozen nothing give rise to all that? Once again, consider Gamow's dolphin. Imagine that back in the early hot days of the dolphin's universe, the vacuum was melted. Whichever way the dolphin swam, everything looked exactly the same. The dolphin wasn't pushed in one direction or the other; it could travel through a little bit of water or a lot, and it still wouldn't be aware of the water's existence.

After a while, this hot universe cooled. The vacuum froze. Assume the dolphin survived, (or a new kind of creature evolved to live in this frozen vacuum). We know that ice is a crystal; it has structure. Traveling through it in a horizontal direction is different from traveling through it in a vertical direction.

The dolphin wouldn't know this because the crystal is the dolphin's vacuum. However, a clever dolphin could make out that his "vacuum" had a kind of invisible underpinning. He'd notice that some kinds of matter behaved differently from other kinds. He might find that if he swam in certain directions he would be repelled or attracted by mysterious forces. These wouldn't be "directions" like compass points, but perhaps directions of "scale," like extremely big or very small. As the dolphin varied his scale, his orientation, his speed, his time, he might find that things would change radically in a way that could only be explained by some invisible underlying structure.

"If he was very smart," said MIT physicist Alan Guth (who wasn't talking about a dolphin but a little person living in a crystal), "he might notice that if he went backward in time, he could see that his crystal was once water at very high temperature. Our current view of physics is that we live in such an ice crystal."

In the melted vacuum, quarks and electrons and gravity and electricity are the same. In the frozen vacuum, they're different. The melted vacuum gives you no way to tell one particle or direction or force from another, so it's a lot closer to "nothing" than a frozen vacuum. Yet the frozen vacuum is the one in which we live.

One question physicists would very much like to answer, in fact, is whether or not the vacuum would freeze in the same way if it had it to do all over again. The consequences of the answer are huge: If the vacuum could only freeze in one way, then everything in our universe is predetermined; if it could have frozen just any old way, then our universe is an accident.

Consider a magnet, which also can "freeze" or "melt" depending on its temperature. A magnet is nothing more than the sum total of the magnetic forces produced by an orderly array of spinning electrons inside a hunk of metal. Each spinning electron creates a magnetic field, like a tiny electromagnet.

When the spins of the electrons all point in the same direction, they pull together, creating clear north and south poles. If you heat a magnet, however, the electrons get jostled around. They point every which way. The magnetic force in one direction gets canceled by an opposite force in the other direction, and the magnetism disappears. The magnetism has, in fact, melted, even though the metal stays intact.

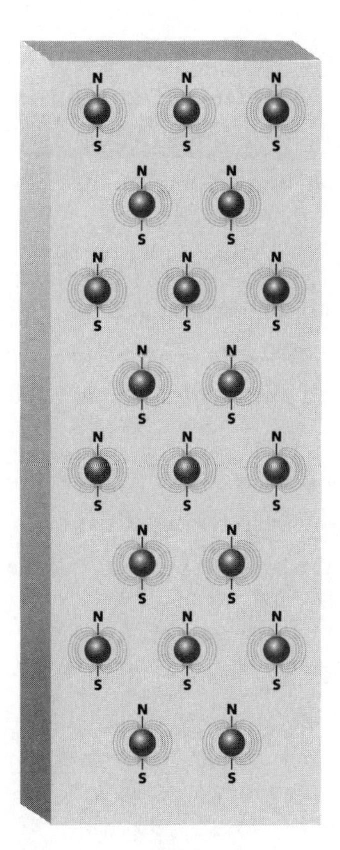

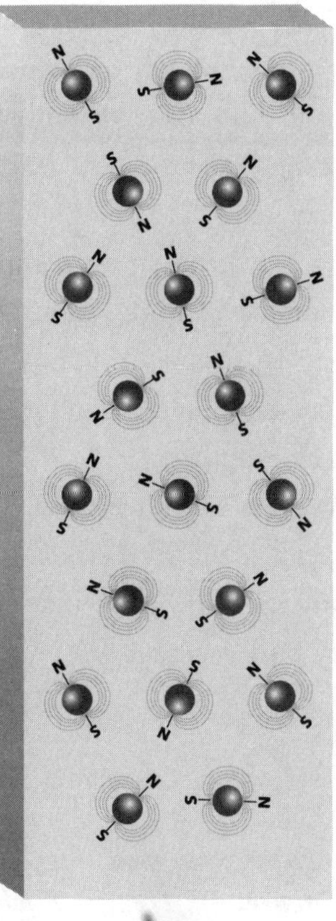

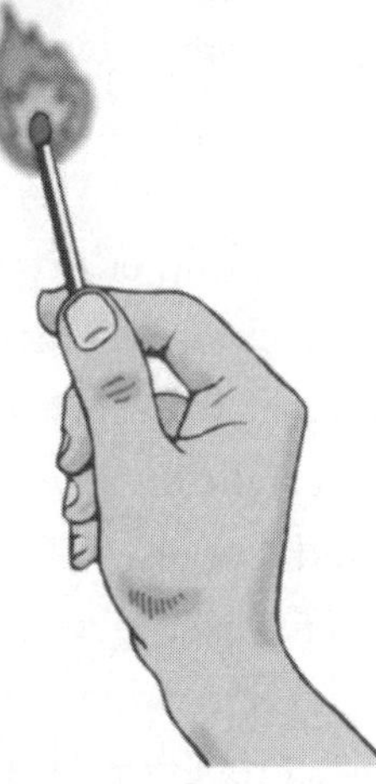

Heat "melts" magnetism. If the magnet "refreezes," the electrons could align in any direction.

When the magnet cools, the electrons will line up again. But not in any particular direction. A few will start pulling together, pulling in a few others, and before long, most of the electrons will be pulling together in a line. But the direction they choose is arbitrary.

This is a bit of an oversimplification, of course. For one thing, the electrons in high-quality store-bought magnets are strongly "persuaded" to line up in one direction or another—which is why north and south poles don't move around every time a magnet gets heated or dropped.

However, it's only natural to wonder whether our vacuum was also "persuaded" by certain inherent characteristics of nature to structure itself in a particular way, or whether it fell into its current state by pure accident.

FREE THE QUARKS

Experiments must therefore "melt" the vacuum to deconfine quarks. . . . The issue is whether we can re-create this early stage of the universe in laboratory experiments. And if we did, would we know that we had?

—*Physics World* magazine

Conceivably, one could answer the question of whether the vacuum had a choice by melting our present vacuum and watching to see whether it freezes back into its current, familiar, form. No one in their right mind would want to try this on a large scale (even if they could) because a successful experiment would destroy the universe.

However, physicists are trying—as you read this—to melt the vacuum in tiny amounts inside huge accelerators that smash atomic nuclei together. If these experiments work, it would be possible to see just what a melted vacuum looks like.

For example, since the structure of the gluons in the vacuum imprisons the quarks, melting the vacuum should free the quarks. Imagine you freeze water, trapping bubbles of air inside; if you melt the ice, the bubbles can escape. The same should be true of the quarks. Instead of quarks trapped in gluon bubbles, physicist should see a clear soup of quarks and gluons wandering around in the melted vacuum.

This won't be easy. For one thing, no one knows what this quark-gluon plasma, as it's called, looks like. For another, it's only possible to sustain such an unnatural state for a small fraction of a second. So the experimenters will have to look not only very carefully, but very quickly.

Roughly speaking, the experimenters are trying to smash the protons and neutrons in the nuclei into such close quarters that the quark bubbles actually overlap, creating one huge bubble with free quarks floating around inside. The vacuum, in effect, would be forced to release its hold on the quarks—at least briefly.*

HUNG UP

It could be that the real universe . . . is perhaps what has been started by some disastrous experiment performed some twenty billion years ago by a post-graduate student in order to test the structure of a vacuum of another universe.

—JOHANN RAFELSKI and BERNDT MULLER, *The Structured Vacuum*

An important question floating around out there along with the free quarks is Why did the vacuum freeze in the first place? And could such a drastic change in the order of things happen again?

*This freely floating soup of quarks and gluons was reported to have appeared briefly in a European laboratory in 1999, but most physicists discounted the account of the "discovery" as premature.

The simple answer is that the vacuum froze because the universe got cold. Water and butter freeze when they get cold, so why shouldn't nothing? But why does anything freeze? Why do loosely connected amorphous water molecules in fluid suddenly fall into formation like a line of soldiers? Why do molecules prefer to pack themselves in well-ordered crystalline patterns just because they happen to be cold?

The reason has to do with energy. Molecules (like everything else in nature) are lazy and don't like to use energy when they don't have to. And it takes less energy to just stand there than it does to move around randomly. As things lose energy and slow down, they fall into whatever state happens to be natural for them: water freezes into crystals, marbles fall to the bottom of bowls, rain falls to earth, and tires fall into potholes.

In the same way, our vacuum froze because it was the easiest thing to do at the time. It fell from a melted state to a frozen one as naturally as apples fall from trees. And it crystallized into the vacuum structure we know in the process. In a sense, the entire history of the vacuum is frozen into its current configuration, a living record of whatever the universe went through in the past.

Could the vacuum fall again? The unsettling answer is yes. Maybe the vacuum isn't at its rock-bottom energy state. Maybe it's just caught on a ledge, like an apple falling from the highest branch of a tree that gets captured in a bird's nest halfway down. The bird's nest isn't really the lowest point the apple could reach. Should a stiff wind come along and knock it out of the nest, it would fall to the ground.

In 1980, Sidney Coleman and Frank De Luccia published a paper that speculated on what might happen if our vacuum suddenly fell from grace. Like an ice crystal forming on a cold window, a tiny patch of *really* frozen vacuum would appear inside our "false" vacuum. In an instant, it would grow exponentially, gobbling up everything around it. The laws of physics would change overnight. (Even faster, actually. And, of course, there would be no day or night.) Among other things, the authors found that this expanding bubble would generate a gravitational force huge enough to crush the universe, a prospect they found "disheartening."

This hasn't happened yet, physicists speculate, because we're still stuck on the ledge, resting on our laurels, thinking we're in a real vacuum, but really existing in a false vacuum that is about as stable as a stick of dynamite lying on a table. It looks tame enough. However, if something comes along and lights it, watch out! The universe could blow up. (As we shall see later, something very like this happened in the early universe, giving rise to everything that exists today, so this notion isn't quite as far-fetched as it might seem.)

After the Coleman and De Luccia paper was published, other physicists realized that they themselves might bring on such a catastrophe with the high-energy "sparks" created in accelerator experiments. Perhaps so much energy would be released in such an encounter that it would blow our vacuum right out of its tree. In fact, it's not impossible that our universe was formed in the first place by an "accident" in another universe (see Chapter 7, "Nothing Becomes Everything").

Luckily, calculations quickly showed that much higher energy particles than physicists could ever create stream from space all the time—and nothing's punctured our bubble yet. Still, this doesn't mean we should necessarily rest easy. "The possibility is not absurd—in our present state of ignorance . . . we would be imprudent to disregard it," writes Sir Martin Rees, one of the physicists who did the calculation. "Indeed, caution should be urged (if not enforced) on experiments that create energy concentrations that may never have occurred naturally. We can only hope that extraterrestrials with greater technical resources, should they exist, are equally cautious!"

A CHIP OFF THE OLD VACUUM

If we could turn the Higgs field off,
we would get a nicer, more vacuumy vacuum.
—FRANK WILCZEK

We have met the vacuum, and it is the quietest possible state of all the fields; fields doing nothing. Sure, they fidget and squirm a bit due to inherent quantum uncertainty, but on average, their value is zero. Or at least, physicists feel fairly safe in calling their value zero.

However, there is one field that is not zero, even in nothing. It never goes away, never fails to assert itself, never settles into its quietest possible state. And because it doesn't go away—can't go away—it's considered part of the vacuum itself. It's the crystal structure that fell into place when the vacuum froze. Even a perfect vacuum, where all other fields wind down to zero, shows a positive value for the so-called Higgs field, named after its inventor, physicist Peter Higgs.

Like every other field, the Higgs field comes in quantum clumps, or Higgs particles. Wilczek likes to call the Higgs particle a "chip off the old vacuum" because it is literally part of the vacuum itself.

How do we know it's there if it's literally nothing? Remember the dolphin living in the frozen vacuum. He knows he's living in a frozen vacuum because things happen in the ice crystal that wouldn't happen if he lived in melted water. The crystal is a part of his vacuum. In the same way, the Higgs field is a part of our vacuum.

"Imagine that the entire universe was permeated with a constant magnetic field," Wilzcek suggests. "You would notice certain things that you couldn't explain unless you assumed the presence of the field. The Higgs field is similar to that situation. We see things that we can't explain, and assuming the presence of the Higgs field explains them."

If you couldn't get rid of this mythical magnetic field, it would be part of what you had left when you took everything else away. In other words, it would be part of the structure of the vacuum.

"There's this ideal vacuum that we can aspire to with the Higgs field turned off," says Wilczek, "but the world we live in seems to have it."

THE IRRESISTIBLE PULL OF NOTHING

Our theoretical physicists call it the Higgs field.
It pervades all space . . . cluttering up your void,
tugging on matter, making it heavy.
—LEON LEDERMAN, explaining the Higgs field
to Democritus, in *The God Particle*

The most critical thing the Higgs field seems to do is give particles (and therefore all matter) mass. What is mass? It's the thing that makes heavy sofas hard to push and bowling balls harder to throw than baseballs. Mass is a measure of inertia—a natural resistance of matter to being pulled or pushed around. Without inertia, everything could travel at the speed of light, always, everywhere.

Alas, most matter can't travel at light speed because it has to slog its way through the Higgs field, which Wilczek also likes to call a kind of "cosmic molasses." This slowing down is what we experience as mass. It's a kind of cross to bear that particles shoulder everywhere they go.

Curiously, not every particle interacts with the Higgs field in the same way. Some pass through without ever paying a toll, just as if it weren't there. Quarks have to slow down and interact with the Higgs field as they pass through, but photons (light particles) do not. Particles that have a free pass through the vacuum travel at light speed. All others slow down.

The more the particle gets bogged down in the Higgs field, the heavier that particle becomes. Therefore, it's the Higgs field that actually gives particles mass. Not only that, the Higgs field determines the mass of each kind of particle—and therefore, in large part, the *identity* of the particle. The Higgs field makes electrons and quarks what they are today—and in turn, makes you what you are today.

Ultimately, the Higgs field might also explain why particles seem to come in families with similar characteristics. It's an odd fact of nature that very similar particles of infinitesimal size vary in mass by huge amounts. Physicists suspect there must be a pattern like the periodic table of elements behind the particle family tree, but no one has been able to find it. Finding such a table would help answer the question: Why are particles the way they are?

Needless to say, physicists would like to get a handle on this little chip off the old vacuum in order to examine its properties up close and personal. They could do this by making such a big tweak in the Higgs field that it would become a real particle. Several large accelerator laboratories—including the Tevatron at Fermilab in Batavia, Illinois, and the Large Electron Positron Collider (LEP) in Europe—are trying to create Higgs particles in high-energy collisions of other particles.*

At even higher energies (probably too high for earthly experimentation), physicists could conceivably even melt the Higgs field. In a melted Higgs field, all particles would have the same mass—which is to say, no mass at all. Everything would move at the speed of light.

In effect, this melting would undo the freezing that took place in the early history of the universe. The energy liberated from the

---

*Until they do create it, the Higgs remains a theoretical entity. It fits neatly into the standard model of particle physics and solves a lot of mysteries, but the concept could still turn out to be wrong.

mass (according to $E = mc^2$) would transform, but not go away. This transformation would be the same as vaporizing someone: Their atoms would disperse. But the material of the body wouldn't disappear from the universe; only the structure of the body would vanish. If physicists melted the Higgs field, they would be vaporizing elementary particles instead of people.

When and if physicists do find (or create) the Higgs particle, they might well discover a whole slew of them—different Higgs particles doing different things in the vacuum. In fact, Lederman complains that so many different particles might make up the vacuum that it's a wonder we can still see the stars though all the muck. The everyday Higgs might be just the "tip of the iceberg," he writes. "A zoo of Higgs boson quanta may be needed to describe the new aether. . . . One longs for a new Einstein who will, in a flash of insight, give us back our lovely nothingness."

Clearly, physicists have good reasons to be nervous about nothing. It's all too complicated for comfort. And so much incessant activity should somehow show up as energy—infinite amounts of energy, in fact.

In the end, there is only one force in the universe that's a neutral judge on matters of energy: gravity. Gravity is a measure of energy, because (according to Einstein) gravity is the curvature of spacetime caused by matter/energy. The next chapter explores in detail how nothing (as well as something) shapes space itself. For now, suffice it to say that according to the scales of gravity, the energy of nothing is nearly nothing. According to the well-understood theoretical calculations, it's everything. Which measurement is correct?

"Clearly the theorists are wrong because the universe is the universe," says Brian Greene.

Clearly, physicists don't quite understand nothing in all its delightful complexity.

If one thing is certain, however, it's that nothing is increasingly difficult to differentiate from something. Particles and forces, matter and energy, stuff and emptiness—everywhere you look, the boundaries are getting fuzzy. Ultimately, the disappearing distinctions between something and nothing may be just another artificial border down the drain.

Chapter 5

NOTHING *BECOMES* CENTER STAGE

There is still a difference between something
and nothing, but it is purely geometrical
and there is nothing behind the geometry.
—MARTIN GARDNER, *The Mathematical Magic Show*

FOR ALL THE FIDGETING of the fields, for all the quantum mechanical vagaries of the vacuum, at least the nothing knew where it stood.

All that bubbling and frothing took place some place, at some time. The perpetually wiggling fields told matter how to move from here to there, but at least there was a here and there to move between. The void itself may have been fickle, but space itself was terra firma underfoot. Time ticked away, monotonous as a metronome. Behind all the to-ing and fro-ing of the vacuum stood fixed reference points that told everything in the universe where it was and when to be there.

The fundamental things applied.

An hour was still an hour, a mile was still a mile.

"Spacetime was out there," says Harvard physicist Andrew Strominger. "You could count on it."

Then, in the early 1900s, Einstein pulled back the curtain to show that static space and fixed time were flimsy facades. Beneath the regular inches and seconds that mark off everyday life, space and time ooze like mud—altering their appearance depending on the motion of whoever happens to be looking.

"Before Einstein, no one talked about changing space and time," says University of Chicago physicist Sean Carroll. But space and time turned to putty in Albert Einstein's hands.

It's hard to overstate the extent of this revolution. It's one thing to poke holes in our notions of nothing; quite another to mess with the time and space in which nothing (and everything) takes place. Probably not since Copernicus knocked Earth off its pedestal in the center of the solar system have scientists so flagrantly pulled the rug out from under people's basic beliefs. But then, as Strominger put it, "The history of physics is the history of giving up cherished ideas."

To be sure, the revolution took place in not-so-baby steps. One step was realizing that space and time were of a piece, woven together into the four-dimensional fabric of spacetime. Another was the discovery that both space and time, individually, are as elastic as bungee cords. It was a further step, still, to see that the fabric of spacetime itself could warp under the influence of matter like hot asphalt under the tires of a heavy truck.

And then, the last straw: Not only could spacetime bend under the influence of matter, it could take matter into its own hands, making things happen. Spacetime ceased to be a dull doormat for the antics of fields. Freed from its bland role as background, the stage became a player, tossing things around in the universe, sucking things up, perhaps even spewing out new universes in the process.

Today, some physicists think space and time may not even be fundamental ingredients of the universe. The concept of nothing, in other words, may extend even beyond space and time to something more, well, empty still. It's hard to imagine, a void without space and time. Where would it be? And when? And yet, some new physics seems to be pushing Einstein's revolution to its logical—if profoundly upsetting—conclusion: If the physicists are right, space and time are toast.

"The real change that's around the corner [is] in the way we think about space and time," says David Gross. "We haven't come to grips with what Einstein taught us. But that's coming. And that will make the world around us seem much stranger than any of us can imagine."

We'll get to this newly emerging strangeness in a later chapter. For now, suffice it to say that even the most far-out physics generally takes place within space and time. Space and time are as close to nothing as physics normally gets. And yet, spacetime is a kind of nothing so bizarre that it makes even the vacuum of empty space seem tame—even boring—by comparison.

NOTHING STICKS

In the purest sense, Einstein showed, space is responsible for bringing down whatever goes up.

—Tom Siegfried, *The Bit and the Pendulum*

Einstein was certainly not the first to think of space as a player. Even Plato wrote that space "sways irregularly in every direction as it is shaken by . . . things, and being set in motion it in turn shakes them." But Einstein turned vague ideas about the nature of space into a precise theory so powerful that it predicted everything from the bending of light by stars to black holes.

At the heart of Einstein's theory—known as the general theory of relativity—is the astonishing discovery that the force we feel as gravity is an illusion. There is no gravitational "glue" holding us to the earth or the earth to the Sun. The only thing that keeps us stuck here on Earth is the sagging of space itself. Gravity is a colossal common misconception. Only the curving of spacetime is real.

How could we have missed such an obvious fact of nature for so long? Easy. For one thing, spacetime warps into a fourth dimension, normally imperceptible to us three-dimensional beings. For another, we're not accustomed to thinking of "forces" as geometry. As we shall see, however, the geometry of nothing may be the cause of everything that is, as well as everything that happens, in the universe.

Since there's no way for us three-dimensional beings to perceive four-dimensional spacetime directly, consider a lower-dimensional analogy: Imagine you're a tiny bug living on a paved street covered with potholes. You walk in and out of the potholes unaware of their existence, just as people walk on the spherical earth blissfully unaware of its curvature. The geometry of the landscape you live in is simply too huge, relative to little you, to be seen as anything but flat. You don't have to think about the potholes in the pavement any more than people in L.A. normally contemplate the undeniable fact that to someone in, say, Sidney, they are dangling upside down.

Even though you don't perceive the potholes, you would notice that trucks and cars repeatedly come to a grinding halt in certain sections of street. You might conclude that some mysterious kind of "force" was pulling them in—perhaps some exotic rubber-attracting magnet.

In the same way, the "force" we feel as gravity is really just the geometry of space—invisible potholes on the road from here to there and now to then. We don't fall "down" so much as we travel through the potholes, following the natural curvature of time and space. Spacetime gets sculpted into various shapes, and the shapes pull us in.

"It almost appears that the physics has been absorbed into the geometry," wrote Sir Arthur Eddington, the astronomer who provided some of the best popular explanations of Einstein's theories. "We did not consciously set out to construct a geometrical theory of the world; we were seeking physical reality by approved methods, and this is what happened."

The geometry of any particular landscape—the configuration of bumps and turns and potholes—is created by whatever matter happens to lie within, just as potholes in the street might initially be created by the pounding pressure of cars and trucks. Heavy objects such as stars warp spacetime just as surely as an elephant sitting on your bed would bend the mattress. At the same time, that very warping changes the paths that other things follow in spacetime. A person sharing a bed with an elephant would be pulled into the "pothole" in the mattress created by the animal's massive weight.

The heavier the object, the deeper the well. The curvature of spacetime is determined only by the amount of matter within it. And since matter is the same stuff as energy, the geometry of spacetime depends on energy as well.

Indeed, if you want to measure how much energy/matter sits in a static region of space, you only need to measure the degree of curvature its presence produced. In fact, this curvature provides the *only* sure way to measure the amount of stuff in space.* Zero stuff means zero curvature, and vice versa. Empty space is flat. Conversely, curved space cannot be empty. If space is curved, then something is sitting in the landscape whether you can actually "see" it or not.

---

*This is true for static space, as inside a box; a different dynamic rules the expanding space of the universe, as we shall see.

"When we perceive that a certain region of the world is empty," writes Eddington, "that is merely the mode in which our senses recognize that it is curved no higher than the first degree. When we perceive that a region contains matter we are recognizing the intrinsic curvature of the world."

This is odder, even, than it seems. Matter is not merely an intruder into the otherwise peaceful landscape of flat empty space. It is no "foreign entity," as Eddington puts it. Rather, the warp in space and matter can be viewed as different aspects of the same thing, like a mountain in the middle of a plain. "The disturbance *is* matter," Eddington writes.

Matter warps spacetime and warped spacetime gives matter and energy weight. It's all wonderfully symmetrical. As in John Wheeler's famous summation: "Spacetime tells matter how to move; matter tells spacetime how to curve."

What gravity does to matter, it also does to space and time itself. The snake eats its tail. Gravity—because it is the curvature of spacetime—affects everything in the universe, including the clocks and rulers used to measure spacetime.

Under the influence of Einstein, time and space changed from isolated and independent clocks and highway markers to infinitely flexible relationships among all things in the cosmos, including space and time. "The relativity theory of physics reduces everything to relations; that is to say, it is structure, not material, which counts," Eddington writes. "The structure cannot be built up without material; but the nature of the material is of no importance."

The very substance of "something," it seems, is written in the geometry of nothing. Flatten out spacetime, and there's nothing to it. Bend it, and something will come.

And lest you fall into the all-too-human habit of thinking of this rubbery fabric of spacetime as something "out there" in the cosmos, remember: Your thoughts, actions, belongings, friends, and family—even the very atoms in your body—are at the tender mercies of this ceaselessly bouncing stage. We live *in* spacetime, rather than on it. This is a hard notion even for physicists to swallow.

"Even if we know better, it is still very common among physicists to think of spacetime as some new kind of object that can be seen from the outside," says physicist Lee Smolin. "On our blackboards and in our notebooks, we draw pictures as if such were the case."

In truth, the imperceptible nothing of space and time is the fabric of our lives: Space and time are us.

"What a long way Faraday's little lines of force have carried us!" exclaims Robert March in *Physics for Poets.* "They started as a way to avoid the problem of action at a distance. Now they generate their own matter. And at the same time, they are the very fabric of spacetime."

NOTHING SHAKES A STAR

What is it that pulls the apple to the ground, bends the circling moon to the earth and makes the planets captives of the sun?... It is intangible time and space themselves, acting in awesome concert as curved space-time holding sway over all things in the universe.

—BANESH HOFFMANN, *Relativity and Its Roots*

Spacetime not only curves; it also vibrates. Like a transatlantic telephone cable, it can carry energy—and messages—from one place to another. If Newton's spacetime were a fixed floor on which everything took place, then Einstein's spacetime would be the head of a huge drum in which everything is embedded. When the drum rolls, vibrations spread throughout the universe. Spacetime not only curves, it moves; it's not only warped, it's dynamic.

"Spacetime is an important part of the action, not just the place where action occurs," says Wheeler, who goes on to question: "Can spacetime be everything—both what is and what happens?" The answer seems to be yes.

For spacetime can warp even when there's no matter around to warp it, even when there's only the energy left over from some long-ago event.

Imagine the Sun sitting in spacetime, bending the local geometry into a bowl-shaped arena (four-dimensional, of course) in which the planets endlessly roll around. If something comes along and shakes the Sun, the shaking Sun sends out ripples in spacetime as surely as a plucked violin string sends out sound waves to our waiting ears. These wrinkles in space persist even when the "plucking" that created them is over.

"Like a giant circus tent in which a single tiny wrinkle can be caused by a tent pole two hundred feet away," writes Wheeler, "space 'here' reflects the influence of 'mass' there."

These waves in spacetime haven't been detected directly only because gravity is so weak. However, they have been seen indirectly by the energy they carry away. In fact, two physicists won the Nobel Prize in 1993 for their discovery that two collapsed stars circling each other in close orbit were losing energy at precisely the rate predicted by Einstein's theory.

Collapsed stars are enormously heavy, so they make big waves in space. That's why physicists were able to measure the energy lost to waves in spacetime by these closely orbiting behemoths. (Actually, the physicists measured the increasing frequency of the light pulses emitted by the stars as they spiraled toward each other—a measure of their increasingly fast mutual orbit.)

But even the doings of lightweight stars can set space in motion, setting up waves that disturb everything in the cosmos—including our home planet. "If you shake a star," says Gross, "the earth will move."

Physicists hope the earth will move enough to detect these ripples in spacetime sometime during the next decade—perhaps even sooner. They've set a clever snare to catch them, called LIGO—for Laser Interferometer Gravitational Wave Observatory. With precisely tuned laser beams bouncing between mirrors set miles apart, even the weak call of gravity should be heard and amplified enough to make the symphony of the stars audible (or at least visible via computer screen) to human ears. Like crashing cymbals, colliding stars or black holes periodically make waves throughout the cosmos. Some of these waves splash ever so gently upon our shores. When LIGO tunes in, we will hear a kind of music never even imagined before—music that is likely to drastically alter our sense of what goes on in the cosmos.

In this sense, looking out into the universe is something like looking into a deep pool that is constantly disturbed by the passage of boats and fishes. Like the songs of the great whales that circle the globe just under the watery surface, the songs of the stars ripple through the heavens. Until now, we have been as deaf to the music of the stars as to the songs of the whales. And yet, it's not that hard to imagine exploding stars and colliding black holes calling to one another through the darkness of space, broadcasting bulletins of their violent births and mergers.

If we're lucky, LIGO and its descendants will help us to hear the ripples in emptiness and read the latest news from the farthest reaches of time and space.

NOTHING GETS REAL

You can't see air, either.

—physicist KIP THORNE, California Institute of Technology

How can you see the curvature of spacetime? Before I count the ways, let me remind the reader that we do not "see" anything directly—not clouds, not houses, not even this book. The image of this page that appears in your mind's eye is nothing more than an insubstantial pattern of neural firings set off by the tweaking of particular proteins in the back of your eye. Those proteins bend and stretch when hit by particles of light.

If you are reading outside, the light particles came from the Sun—a journey of about eight minutes from the star's blinding surface. These visible photons hit the page at light speed, making only the shortest rest stop on a patch of black or white before bouncing into your pupil. They then fall on your retina like rain, mixed in with countless other photons—those bouncing off your nose, for example, or the grass outside; those traveling through the thick transparent goop that keeps your eye in shape (sometimes marred with tiny black threads we see as "floaters"); those from the blood vessels in your eyes and the various stray reflections from leaf, sky, dog, fingers, table, wineglass.

All these photons together create an image on the back of your eye that is a mishmash of misinformation: upside-down, distorted, confused, and woefully incomplete. And yet, you "see" the words printed on the page.

Of course, you don't really see the words at all. You see the light. The words are something that happened to the light between the Sun and you. Your brain interprets the distortions in the sunlight as words.

In the same way, physicists "see" the curvature of space by what it does to intervening objects. They see curved space by what it does to light in the same way that you see words on a page.

In fact, Einstein's theory of gravity as curved space was first proved when light skimming the sun during a solar eclipse was seen to bend. The curvature of space was "seen" because the light was displaced. Light bending through curved space behaves almost—but not exactly—like light traveling through a glass lens. Light passing through a lens bends, but the lens itself remains rigid. Light passing through curved spacetime, on the other hand, does not, strictly speaking, bend; it simply follows the shortest possible path. The spacetime itself bends, taking the light with it.

This means that the bending of light as it passes through the warped space near a star is not merely *evidence* that space is curved; it *is* the curvature of spacetime. This is central: If gravity were only "pulling on starlight," then good old Newtonian ballistics would explain most of the bending we see. Einstein showed that gravity is the property of the background, not of objects within it.

Over the past decade, astronomers have become adept at making use of this bending to see various objects in the universe, much as they use the bending of light through a telescope lens to view otherwise invisible stars. So routine has this practice become, in fact, that it's hard to remember when a "gravity lens" was merely a wild glimmer in Einstein's eye. Recent deep-sky images from the Hubble Space Telescope reveal a landscape littered with tiny bright crescents, like so many cosmic fingernail shavings. Each is a galaxy, far, far away. Viewed edge-on, a distant galaxy appears as a tiny bright line. But viewed through severely distorted space, it curls like a lemon peel.

The curling tells us that it's a bumpy road from there to here—clear-space turbulence, you might call it.

These same gravity lenses can create multiple images of galaxies, making a single galaxy into a pair, or even a group, of identical twins or, perhaps, quadruplets. Gravity lenses can also focus light from distant galaxies into a bright point of light, like sunlight passing through a magnifying glass. This focusing effect has already led to the discovery of several otherwise invisible dark stars that may make up a good portion of the mysterious missing matter in the galaxy.

And curved space does more than bend light; it's also palpable. Since the curvature of space is gravity, you measure it every time you step on a scale.

"Change the shape of space and one immediately changes perceptions," says philosopher Graham Nerlich in *The Shape of Space,* a book dedicated to proving, quite successfully, that empty space is as real as the objects within it. Extremely curved space would form a hole composed of nothing but empty space, he points out. If we encountered such holes, "we could palpate their contours and ache with the pressures of keeping our hands in the parts of deepest curvature. If that does not show that the hole is concrete, what could possibly count as concrete?"

In a sense, the earthly tides are merely the way the oceans perceive the shape of space: Twice a day, the curvature in space caused by the nearby mass of the Moon pulls the oceans out of shape, causing the water to rise and fall upon the shores.

If you fell down a black hole, the tidal forces would be so immense that they would stretch your body into a string of infinitely thin spaghetti.

DIGGING A BLACK HOLE

Of all the conceptions of the human mind, from
unicorns to gargoyles to the hydrogen bomb,
the most fantastic, perhaps, is the black hole.

—KIP THORNE, *Black Holes and Time Warps: Einstein's Outrageous Legacy*

If you are still not convinced that spacetime is a real "something," consider some of the objects created from pure spacetime and nothing else. Among these, nothing quite stands up to the notorious black hole, so named because its colossal gravity sucks up everything, including light.

A black hole is born, astronomers believe, when a massive star burns up all its nuclear fuel. With nothing left to push "out" against gravity, the star finally collapses under its own weight. As the star collapses in on itself, its mass gets concentrated in an increasingly smaller space. Ultimately, the entire star gets squeezed into a single dimensionless point of zero dimension, zero volume. The resulting curvature of space is so severe that anything in the vicinity gets trapped inside, light and matter alike. Anything that falls into the hole is lost forever in space. Everything goes in, nothing comes out.

"For all practical purposes," says Thorne, "[the star] is completely gone from our universe. The only thing left behind is its intense gravitational pull."

Like Berkeley's "ghosts of departed quantitics," a black hole is but a memory of what has come and gone away, a mere trace of its former self. Along with the hole goes everything that ever fell into it—not only the whole star, but also information about what made up the star in the first place, as well as anything (like a planet, for example) that may have been attached to it.

"The star is, for all intents and purposes, plucked out of spacetime," writes Robert March.

Unlike unicorns, however, black holes should be quite common in the universe at large. In fact, UCLA astronomer Andrea Ghez and colleagues have measured the tight orbits of stars twirling about the center of our own Milky Way so precisely that astronomers are all but certain a massive black hole lurks there, holding our galaxy together. For physicist James Hartle, who directs the Institute for Theoretical Physics at UC Santa Barbara, these measurements say black hole loud and clear. "Black holes are no longer a theorist's dream," he said of Ghez's work at a recent physics meeting. "They have been detected."

With so many black holes out there, these massive objects should be running into one another all the time, colliding and merging into even more impressive puckers in spacetime. It is the wake from such a collision, in fact, that LIGO is primarily designed to detect.

And if Britain's Astronomer Royal, Sir Martin Rees, is right, it's probably a good idea to have gravity-wave detectors on the lookout: In his book, *Before the Beginning,* Rees suggests that a newly merged black hole could recoil from the collision that created it with such a kick that it would be thrown out of its galaxy. "Such massive holes . . . could now be hurtling through intergalactic space," he says. Indeed, in late 1999 NASA's Hubble Space Telescope found evidence of the first lone black holes adrift in our galaxy.

A HOME FOR ODD SOCKS

Kip will come back as Hawking radiation.

—STEPHEN HAWKING, on the occasion of his friend Kip Thorne's sixtieth birthday, explaining why even if Thorne jumped down a black hole, he wouldn't be lost to the universe forever.

Up until the 1970s, everyone thought that a black hole was a one-way street to oblivion. But British physicist Stephen Hawking showed that severely warped spacetime could be a lot more complicated than anybody thought—especially when quantum mechanics was stirred into the mix.

As it turns out, black holes not only swallow stars, effectively turning something into nothing, they also *create* something out of nothing. Eventually, certain kinds of black holes could spew out so much "something" that they should evaporate away.

This odd behavior depends on the fact that those virtual particles bubbling out of the vacuum are only as real as the virtual image of yourself behind the mirror. However, they can become real if suffused with enough energy. And as Hawking figured out, the severely curved space at the edge of a black hole provides just the right amount of punch to turn a virtual particle into a real one.

Of course, particles created from nothing—virtual or not—are always created in pairs, one part matter to an equal part antimatter. In Hawking's scenario, however, one of the pair gets sucked down the hole, while the other flies off into space. To an observer outside of the hole, it looks as if the hole is "radiating" energy into space. And just as the radiant energy of the Sun makes it shrink ever so slightly with each passing second, so black hole radiation causes certain kinds of black holes to melt completely away.

Before Hawking came up with black hole evaporation, there was hope that information lost in black holes was somehow stored and could—at least theoretically—be retrieved. However, if the hole evaporates, then anything in the recesses of the hole is destroyed forever. In go stars, houses, computers, whatever. Out comes amorphous radiation. The information can't return to the universe carried on the particles created at the edge of the hole because these particles are created just outside the hole and have no real contact with whatever fell in.

A black hole just sitting there, in other words, merely loses information; a black hole that radiates destroys information forever. Evaporating black holes turn information to mush.

"If one takes Einstein's general relativity seriously," Hawking concludes, "one must allow for the possibility that spacetime ties itself in knots and that information gets lost in the folds. . . . Maybe that is where all those odd socks went."

To a physicist, this sets off alarm bells of the most serious kind. If what goes into the hole isn't connected with what comes out, then the future is severed from the past—an unholy act if ever there was one.*

*Luckily, many physicists now believe that the information does not get lost, but is somehow stored at the event horizon. This is connected with Susskind's work on holographic information storage, described in Chapter 6, "Nothing Gets Strung Out."

NAKED?

Every black hole brings an end to time
and space and the laws of physics.
—JOHN A. WHEELER,
Geons, Black Holes, and Quantum Foam

Something even more ominous takes place in the heart of black holes: they put an end to space and time. In the pinched-off centers of black holes, spacetime appears to simply stop. This is all the more surprising since black holes *are* space and time, albeit severely curved. Yet there it is: Every black hole dead-ends in a literal point of no return, that zero-dimensional object called the singularity. At the singularity, space and time themselves get torn to shreds (more on this later). "The singularity acts like an edge," says Gary Horowitz. "You run into it, and it's the end. There's no time after that; there's no space after that."

Many physicists believe that nature has contrived to shield us from the sight of such shameless horrors. Every black hole is surrounded by a boundary called the event horizon. Approaching the hole from outside, you could still escape its clutches so long as you don't cross this invisible line. Beyond that point, however, you are doomed. On the highway to a black hole, the event horizon is the last exit.

Since light can't escape from inside the event horizon, information about the horrors inside can't make it to the outside world. The hole is never seen "naked"; rather, it is always "decently hidden," as Hawking puts it. Mathematician Roger Penrose believes that the cloaking of singularities is an unwritten law of nature that can never be broken, no matter what. His name for this unwritten rule is "cosmic censorship."

But no one has been able to prove that cosmic censorship must be in force at all places and times. For example, Hawking suggests that an exploding black hole might offer the briefest glimpse of the forbidden territory deep inside. Since physicists believe that the big bang is, in effect, an exploding black hole, the birth of the universe itself might be that dreaded "naked" singularity. As Hawking puts it: "Cosmic censorship may shield us from black hole singularities but we see the big bang in full frontal nakedness."

Fortunately, most black holes would still be "protected" from our gaze by the iron curtain of an event horizon. That's a good thing, as we shall see, because when the laws of physics break down, anything—literally—can happen. "Science fiction films always dramatize the event horizon as a sort of cosmic Venus flytrap," says astronomer John Barrow. "Its real importance is as a shield against what would otherwise come out into the Universe."

YOUR MOTHER'S A BLACK HOLE

{E}ach black hole is a bud that leads to a new universe of moments.
—LEE SMOLIN, *The Life of the Cosmos*

Black holes have a creative side, as well: for example, they may be thresholds to other universes, or the seeds of new universes, depending on how you look at it. According to some theories, everything that goes into a black hole may tunnel through to another universe in something called a white hole—a black hole's mirror image. The singularity punctures spacetime and opens a porthole to another world. (We'll return to the subject of how universes can be created from singularities in Chapter 7, "Nothing Becomes Everything.")

Lee Smolin has taken this line of argument one unsettling step further. Since universes with many stars collapsing into black holes would also spawn great numbers of new universes, you might say that universes full of stars produce more "progeny" than other universes. In a version of Darwinian selection, these "fittest" universes are also the most likely to give rise to life, because life depends on stars. That would make us the result of an evolutionary competition for procreativity among—of all unlikely things—black holes.

A black hole, in other words, might be your primordial mother.

Whether this is true or not, black holes remain sources of endless fascination. They come in all sizes, from the submicrocosmos of particles to the vast expanse of galaxies; they can be tiny portholes or huge chasms.

A black hole with two ends that puncture two different points in spacetime could conceivably provide rapid-transport systems to other times and places. Kip Thorne—who created the wormhole transport system used by Carl Sagan in his science-fiction novel *Contact*—is one of several physicists who have seriously considered the possibility of time machines constructed out of wormholes. In effect, one would travel from one place to another through nothing at all.

Unfortunately, all schemes for making time machines have so far proved unfeasible. For one thing, keeping such a wormhole open would require "negative energy," and it isn't clear that negative energy exists or, if it does, that it can be kept around for long enough to support the travel of real persons. If negative energy did exist, of course, it could be taken to mean that time machines would be made of even *less* than nothing.

For another, quantum mechanics introduces an unexpected disaster into the time machine scheme. Thorne and his colleagues showed that if a wormhole was used as a time machine, subatomic fluctuations would pile up on one another inside the wormhole and explode. The time machine would self-destruct.

And even if the time machine could survive the explosion, it would pose conundrums for cause and effect. If you could travel back in time, you could murder your grandfather, and then you wouldn't exist. Like Hollywood scriptwriters, Thorne and other physicists have explored various scenarios for getting around this problem. "It is not hopeless," says Thorne, "but I'd give heavy odds that explosions destroy all time machines, so we needn't face the conundrums."

Thorne's colleague Hawking shares his pessimism. Hawking believes that the universe protects itself against time travel with a "chronology protection" mechanism that "keeps the world safe for historians," as he likes to put it. Hawking has also said that the best experimental proof that travel back in time isn't possible is the noticeable absence of hordes of tourists from the future.

ANOTHER DIMENSION TO NOTHING

There is a fifth dimension,
beyond that which is known to man.
—ROD SERLING, *The Twilight Zone*

As the reader may have noticed, the blank slate on which everything takes place has grown to four dimensions. Since this story line is only going to get more important (and more tangled) as we proceed, we might pause for a moment to answer an obvious question: Where is this extra dimension? What is an extra dimension, anyway?

In the simplest sense, a dimension is a way you can move, which is why mathematicians call it a "degree of freedom." Normally, you can move in three perpendicular dimensions—thus the three-dimensional world you live in (plus one of time). You can go east-west, north-south, or up-down, or any combination in between.

Equally, you could think of dimensions as coordinates. On a two-dimensional surface, such as the surface of the earth, any position can be exactly defined with only two coordinates: latitude and longitude. For positions above or below sea level, a third coordinate is needed—altitude. To specify a time, add a fourth coordinate.

When you agree to meet a friend at 2 P.M. at the corner of Hollywood and Vine on the twelfth floor, you require four dimensions to pin down your location.

More important, the number of dimensions determines what's possible in a universe and what isn't. Consider, for example, the life of a two-dimensional comic strip character, Cathy. If Cathy is stuck on a bad date, she cannot move out of her frame. She can move vertically in the strip, or horizontally, but she cannot escape the two-dimensional surface of the paper to scale the walls of her cartoon world. Such a move into the third dimension would be literally out of her universe—as unthinkable to Cathy as a fourth spatial dimension is to us.

However, if Cathy could leap over the comic strip frame, she would be free. Going to a higher dimension presents opportunities because there's a lot more room to move around.

The same would be true if Cathy were a real three-dimensional person trapped in a three-dimensional house. Again, she could use a higher dimension to get out of her fix. By traveling backward in the dimension of time, she could move back in history to a time before the house was built. Presto, she would be free!

And so on, up to higher and higher dimensions. You can leap over a four-dimensional barrier in five-dimensional space, untie an eight-dimensional knot in nine-dimensional space. Magical things are made possible by a mere change in the dimensions of space.

This explains, perhaps, why religious miracles have been attributed to goings-on in the fourth dimension. In the seventeenth century, philosopher Henry More even suggested that the fourth dimension was the habitat of ghosts.

But these different dimensions also have real, physical, properties. There is an old riddle that illustrates this well: A person walks one mile due south, then one mile due east, then one mile due north, and winds up back exactly where she started. Question: What color are the bears?

The answer is supposed to be that the bears are white, because the person is at the North Pole—the only place on earth where one could travel such a triangle.* But the larger point is that three right angles on a spherical surface (the earth) can take you back where you started. On a flat surface, like a small piece of paper, it takes four right angles (a square). So geometry not only shapes space, the dimensions of space shape geometry.

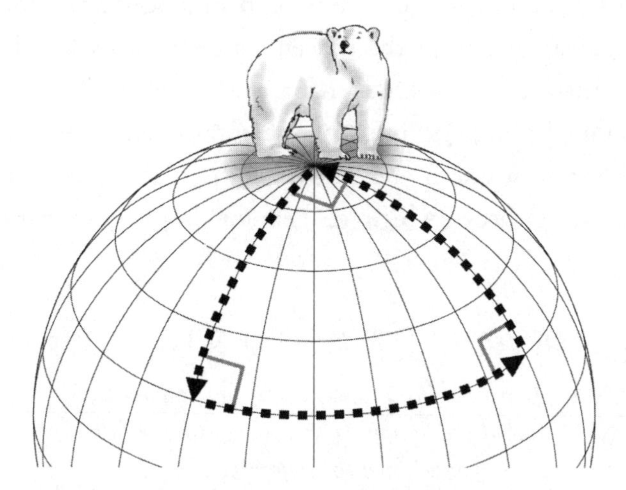

Traveling due south, due east, and due north gets you back to your starting place at the North Pole.

This is true for any set of dimensions, not just the two-dimensional surface of a sphere. In four-dimensional spacetime, the angles in a triangle can add up to more or less 180 degrees, depending on precisely how the space is curved. Conversely, by measuring the angles, astronomers can deduce the geometry. As it turns out, the curvature of four-dimensional spacetime has been measured

*This simple riddle hides a lot of interesting complications. For one thing, the North Pole is obviously not the only place on earth where one could draw a triangle with three right angles; after all, the earth is a sphere, so geometrically speaking, any starting point is the same as any other.

quite well both locally and on the large scale of the universe. Recently, astronomers measuring radio waves from distant quasars in a huge cosmic triangle confirmed Einstein's spacetime curvature to an accuracy of 3 parts in 10,000, according to Washington University astronomer Clifford Will.

Many physicists—including Einstein—have speculated about the possibility of adding a fifth or even more dimensions to the four-dimensional framework of spacetime to explain yet other forces. In 1919, Polish mathematician Theodor Kaluza proposed that electromagnetism was due to the warping into an unseen fifth dimension, just as gravity is due to the unseen warping into a fourth. Then Swedish mathematician Oskar Klein suggested that this extra dimension might be coiled into tiny subatomic-scale tubes.

In a sense, Kaluza's and Klein's ideas have been resurrected today in string theory, which we'll explore in the next chapter.

YOU CAN'T LEAVE HOME WITHOUT IT

It reaches inside every candy box, no matter what the wrapping, to distinguish the full pound from the empty container.

—Philip Morrison and Phylis Morrison, *Powers of Ten*

Gravity's unique position as the geometry of spacetime itself makes it very unlike other forces. You can't cancel it out or get away from it. Gravity reaches inside your clothes to tug on your underwear, inside your skin to tug on your heart, inside the earth to tug on the iron core. It brings together every corner of the universe. No other known force is so relentless. Gravity pulls on space and time, like a dog chasing its tail. Meanwhile, everything that pops in and out of the vacuum warps the space around it, creating a gravitational field. This includes gravitons—or particles of gravity—which in turn create space-and-time warps around themselves. This situation is self-referential in the extreme, like a statement that reads: "This sentence contains five words."

Because gravity reaches everywhere and everything, it cannot be excluded from the messy subatomic world. It gets tied up in the vacuum with all its frothing, fuzzy chaos. Where huge amounts of gravity are squeezed into close quarters, the vacuum and spacetime collide in utter turmoil. The roiling vacuum "unglues space and time," says Thorne, and then "destroys time as a concept. . . . Space, the sole remaining remnant of what was once a unified spacetime, becomes a random, probabilistic froth, like soapsuds." In such a state, gravity becomes not only circular, but paradoxical, like a statement that reads: "This sentence is false."

The smoothly undulating landscape of gravity gets chopped up into a mass of disconnected bits that are tossed about like so much flotsam riding the quantum mechanical fluctuations of the vacuum. Einstein's lovely fabric of curved spacetime gets pulled into the quantum mechanical equivalent of a Cuisinart. "When gravitational fields become strong," as Alan Guth puts it, "space begins to behave more like quicksand."

For this reason, Wheeler dubbed this state of affairs "quantum foam" to distinguish it from Einstein's softly undulating fabric. In these foamy depths, you can forget about normal ideas like dimension altogether. There is no left or right, no before and after. "Ordinary ideas of length would disappear," writes Wheeler. "Ordinary ideas of time would evaporate." Only "the roiling chaos of weird spacetime geometries" would be left.

One obvious place where you might find such a state of affairs is the singularity at the bottom of a black hole—a place where "space can be crumpled like a piece of paper," as Wheeler describes it, and "time can be extinguished like a blown-out flame."

Another is the big bang itself. As we dial back the clock of the universe to the beginning of time, the universe gets smaller, gravity gets stronger, and quantum uncertainties grow larger. Finally, the uncertainty becomes larger than the time interval itself. It's as if the tremble of your hand as you measure the diameter of a snowflake is larger than the snowflake itself.

Time at the first moment dissolves into nonsense.

"The whole question 'When is it?' becomes meaningless," said Sidney Coleman. "If you ask questions about what happened at very early times, and you compute the answer, the [real] answer is 'That's not the right question.' Time doesn't mean anything."

Alas, calling such a confused state of spacetime "foam" doesn't tell us exactly what's going on. A new term, MIT physicist Victor Weisskopf used to like to say, is like an empty suitcase: It doesn't take you on a very long trip. Calling it foam "is not telling you what it is," says Andrew Strominger. "It's like saying, 'A lot of stuff is happening and we don't understand it.'"

A MISMATCH MADE IN HEAVEN

[G]eneral relativity and quantum mechanics, when combined, begin to shake, rattle, and gush with steam like a red-lined automobile.

—BRIAN GREENE, *The Elegant Universe*

The source of the problem (as well as the foam) is this: Einstein's theory of gravity (general relativity) perfectly describes everything that happens on large scales in the cosmos. Quantum mechanics perfectly describes everything else. Both sets of laws have survived decades of stringent experimental testing. Quantum mechanics gave rise to lasers and computers; Einstein's theory of gravity predicts everything from the shape of the universe to the fast-running clocks on GPS satellites. But putting the two together leads inevitably to irreconcilable differences.

While physicists argue over which theory will ultimately have to bend (or break down) to mend the marriage, it is hard to see how either could accommodate much change. "It has been said that quantum field theory is the most accurate physical theory ever, being accurate to about one part in 10^{11}," writes Roger Penrose, in *The Nature of Space and Time.* "However, I would like to point out that general relativity has, in a clear sense, now been tested to be correct to one part in 10^{14} (and this accuracy has apparently been limited merely by the accuracy of clocks on earth)."

And yet, it's clear that one—or both—must be wrong, because the two theories are mutually exclusive.

Consider the Sun—an object central to human life. The Sun provides all our heat and light, both of which obey the laws of quantum mechanics. But the Sun also provides gravity that keeps Earth in orbit. It curves the space around it into a "gravity well" that keeps us circling, once a year, around the center.

"What is the Sun?" Sean Carroll, physicist at the University of Chicago, asks rhetorically. "A source of gravity? Or a source of heat?" Current theories won't let us have it both ways. The two faces of the Sun speak in entirely different languages, with no known key to translate between them. The mathematics—as well as the physical ideas behind the math—are entirely incompatible.

In the language of gravity, the Sun's quantum mechanical aspects are pure gobbledygook. And vice versa. "We can describe the world that we see and experience completely," says Carroll, "but the explanations are internally inconsistent." The gently rolling landscapes of Einstein's spacetime simply can't live within the same framework as quantum mechanics' probabilistic froth.

In our everyday world, such cosmic inconsistencies have no seeming effect. Quantum uncertainty blurs only things as small as atoms, not everyday objects like chairs; Einstein's warpage of space and time shapes the orbits of planets but is too dilute to be seen on the scale of our own backyards.

Where the large-scale fabric of spacetime gets tangled in the inner lives of atoms, however, space and time fail to make sense.

GEOMETRY IS DESTINY

I have . . . again perpetrated something about gravitation theory which somewhat exposes me to the danger of being confined in a madhouse.
—ALBERT EINSTEIN

An apple warps spacetime ever so slightly. The earth warps space strongly enough to hold on to apples, the atmosphere, and even the Moon. The Sun's warp creates a well so deep that none of the nine planets can escape. The stars in a galaxy all contribute to the sagging of space around them, so that all are pulled together into a glob, ellipse, or spiral.

But Einstein's conclusion that matter determined the overall shape of the universe at large surprised everyone. Einstein himself was so impressed by the magnitude of this conclusion that he joked to a friend that people might want to confine him to a madhouse merely for suggesting it.

Nonetheless, this conclusion follows naturally from Einstein's theory of gravity: The amount of matter/energy in the universe determines its geometry. Perhaps even more frightening, it determines our ultimate fate. A universe with enough matter will eventually gravitate in on itself, collapsing like a burned-out star, perhaps to explode again into a new universe. A universe short on matter, in contrast, will not hold together over the long run but will dissipate forever into the darkness.

Gravity curls up spacetime. That is a fact. The question is only How? And the answer depends on how much matter and energy are out there. In standard parlance, the alternatives for the shape of the universe are "closed," "open," or "flat": The universe is closed (like a sphere) if it has enough matter to curl up; open (like a horn) if it has too little to hold together; flat if it has just the right amount of matter to balance precariously in between.

Of course, cosmologists know a good deal more about the universe at large than they did when Einstein first came up with his theory. (In fact, there weren't any cosmologists around then, because the study of the universe at large wasn't really a science.) Einstein's equations relating matter and energy to the curvature of spacetime allowed space to expand or contract, but not to stand still. Since everyone at the time assumed that the universe pretty much just sat there, Einstein introduced a soon-to-be-famous "fudge factor" to make the universe behave according to conventional wisdom. (Actually, he had better reasons as well.)

Astronomers soon discovered, however, that the universe does, indeed, expand, and it has been expanding ever since the big bang. In a sense, the big bang is still going on, albeit at a quite sedate pace. We are still riding along with the expansion that started some 13 billion years ago, when everything presumably exploded from a single point in space and time.

But the overall shape of the universe is still something of a mystery. For all intents and purposes, it appears to be perfectly flat. That doesn't mean it's empty, but rather that the inward pull of gravity on all its parts is exactly balanced by its outward expansion. In other words, the amount of curvature isn't zero, but *net* zero. Gravity pulls matter in; expansion pushes it out; the result is a flat line.

The reader might well ask at this point: How do scientists know the universe is flat? The most definite answer comes from recent measurements of the leftover glow of the big bang, the cosmic microwave background, mentioned earlier. Because this glow carries the imprint of the very early universe, it carries information about its shape. University of Chicago physicist Michael Turner calls the cosmic microwave background a "geometry meter" that can accurately measure the geometry of the universe. The answer seems to be that it's flat.

But even before the background glow was measured, physicists had good reasons for believing that the universe was flat: A universe with enough matter to curl in on itself would have collapsed under its own weight long ago; conversely, a universe with too little matter to ultimately hold together would have dissipated before stars or planets could have formed. Only because the universe did neither do we exist at all to ponder the question.

"The universe gives every sign of performing this balancing act," said University of Chicago cosmologist Rocky Kolb. "It's as if magically the taxes you take in balance the budget to the penny."

The problem is, there doesn't seem to be enough matter in the universe to even the scales. Astronomers who "weigh" the matter in the cosmos by its gravitational pull come up substantially short; so do theorists who cook up matter out of equations from scratch.

In fact, visible matter—stars, planets, gases, and the like—only makes up about 1 percent of the necessary matter to explain this unreasonable flatness. Astronomers assume the rest is made up of some kind of dark matter, not yet seen. Of this, only a small percentage is the normal variety of matter that makes up planets, stars, us. The rest is unknown. The only sure thing is it's not like us. In fact, the puny 5 to 8 percent of matter that *is* like us has to follow the gravitational pull of the far more powerful dark matter like a puppy on a leash.

The variety of possible candidates for this dark matter is enormous: the theorists' creations range from particles billions of times more massive than the ordinary protons in atoms to particles billions of times lighter than electrons. Some—the so-called VAMPS, or variable mass particles—even change their properties over time and space. Candidates are as familiar as well-known particles like neutrinos—but with a small amount of mass—to far more speculative notions such as shadow matter and even wimpzillas. "It is a remarkable range of ignorance," University of Washington astronomer Christopher Stubbs recently pointed out. "We really don't know what's going on."

But then, perhaps it's not so surprising that the nature of the dark matter is elusive. We still don't know the reasons behind the properties of the everyday matter we *can* see and weigh and subject to experiment. "Since we have no idea why the particles we see have the masses they do," writes physicist Lawrence Krauss in *Quintessence,* "we know even less about the particles we have not seen."

The bottom line, however, is that most of the matter that determines the shape of the universe is dark—and this moves dark matter to the center stage of physicists' concerns. "There is unquestionable evidence," writes Krauss, "that the formation of everything we see is governed by that which we now cannot."

Not surprisingly, physicists have stationed a wide variety of detectors around the earth (as well as under it and above it) to catch a particle of dark matter.

In addition, several new satellites will be launched to view the big bang's afterglow in far better detail, and some physicists expect they might be in for more surprises. For example, on very large scales, is it possible that the universe is curved not into a sphere but, say, a doughnut? What about a two-holed doughnut? Einstein's relativity can predict whether the universe curls inward or outward depending on the total amount of energy/matter that lies within. But his theory cannot say anything about the overall topology—that is, whether there are holes or twists. If the universe is shaped like such a doughnut, it would look like a hall of mirrors; as we look at the sky, we should see multiple images of ourselves. Only time, and the microwave background, will tell.

A WRINKLED PEA?

There is a whole family of pea instantons.
—STEPHEN HAWKING, speaking at Caltech, March 1998

While some physicists look into the future or out into space to find out about the overall shape of spacetime, others look back to its origins. In some versions of cosmic history, the primordial speck of nothing that spawned the universe looks like a figure eight or barbells; Stephen Hawking has suggested it looks like a wrinkled pea. He also came up with a way to have it look like a closed sphere in certain dimensions, an open parabola in others, depending on how you slice it.

Until a few years ago, Hawking firmly believed that the universe was closed; in other words, the fabric of spacetime was shaped in something like a sphere, only in four dimensions. It has a beginning and an end—just as the earth has a north and south pole—but no boundaries or edges. Hawking likes this kind of universe because it avoids the ugly big bang singularity that breaks up space, time, and the laws of physics.

But recently, he was forced to acknowledge that the real universe may be open, after all, and curve outward like a horn. (See why in Chapter 7, "Nothing Becomes Everything"). No problem. By going to new dimensions, Hawking figured out a way to have it opened and closed simultaneously. In eight-dimensional space, the universe can be both, depending on how you carve it up.

Imagine spacetime as an ice-cream cone. If you slice it horizontally, the section you carve is a circle—a closed universe. If you slice it top to bottom, you can get a parabola—an open universe.

And the pea?

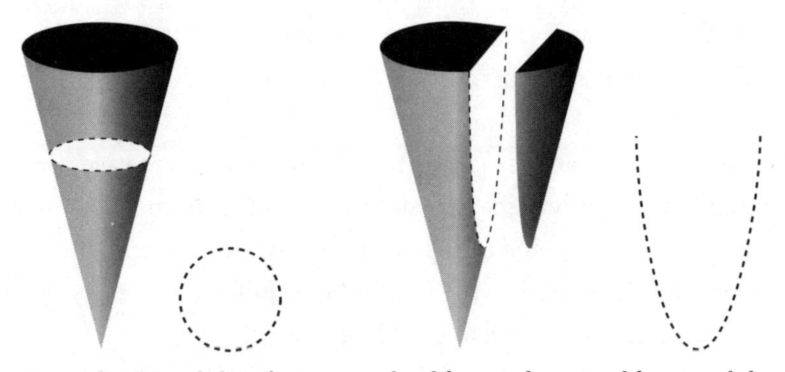

A two-dimensional slice of a cone can close like a circle or open like a parabola.

Hawking proposed that this open and shut universe evolves from a speck of nothingness he calls the "pea instanton." An instanton is a particle of space and time. It is not so much a "thing" as an "event"—a dollop of nothing that transforms into the beginning of everything. Since this instanton would not be perfectly spherical, but rather wrinkled up, Hawking called it a pea. He told his mostly puzzled audience at Caltech that there would be a whole family of pea instantons.

Still, Hawking has a way of being right even when he's most outrageous. As Caltech physicist John Preskill explained after the lecture, "He has a feel for what is the right answer."

TWISTED, BEADED, BRAIDED, KNOTTED, POLKA-DOTTED

Everything you see in the universe, microwave fluctuations, clusters, galaxies, stars, planets, people, poodles, pond scum, Linda Tripp, everything comes from {those} fluctuations.

—ROCKY KOLB

The patterns written into the blank background of space and time are many, rich, and deep. While this entire book could probably be devoted to describing them, I'll only mention two more—rather remarkable—shapes that spacetime can assume. Both harken back to the quantum mechanical vacuum fluctuations described in the last chapter.

One of these patterns is clearly written in the microwave background; in fact, it was the first sighting of this pattern that caused at least one astronomer to proclaim he had seen what a religious person might describe as the face of God. But the real story is in how the pattern got there in the first place.

Take a hunk of space and time, of nothing, the primordial instanton. Let quantum mechanical fluctuations set it all aquiver. Then let the instanton expand into all the universe. What happens to those tiny subatomic-scale wiggles? They puff up to enormous proportions along with the universe. Today, they are as big as clusters of galaxies. In fact, these tiny early fluctuations, physicists believe, are behind all large-scale structure in the universe. They formed the seeds for everything that exists today.

"The intricate patterns of galaxies and clusters of galaxies may be the product of quantum processes in the early universe," writes Alan Guth. "The same . . . uncertainty principle that governs the behavior of electrons and quarks may also be responsible for Andromeda and the Great Wall."* And because these fluctuations were the seeds that allowed stars to form, they are also the progenitors of every element produced in stars, which includes everything except hydrogen and helium. That means everyone who has ever lived carries the imprint of these early twitches in their very atoms.

These fluctuations also imprinted the first light from the universe, the afterglow of the big bang, which is what allows them to be seen. Soon, astronomers will be able to study the messages they carry even more closely.

---

*This is not the Great Wall of China, but a vast string of galaxy clusters in the sky.

Other wrinkles in the background are, for now at least, simply (intriguing) theoretical possibilities.

Sidney Coleman, for example, has proposed that spacetime might be riddled with a complex system of wormholes and bubbles on a very small scale, and that this wormhole system might account for much of the puzzling "weight" of empty space. Somehow, the extra weight leaks out through the wormholes into other universes.

Other structures might be more like defects or knots in the warp and woof of the fabric of spacetime itself. These structures might have been etched into spacetime when the early vacuum froze. Just as freezing water does not usually form perfectly uniform crystals, the freezing vacuum would have been riddled with irregularities. This freezing, physicists believe, might have spread in rapidly advancing sheets. Where the sheets met, boundaries would have formed, perhaps trapping bits of the primordial vacuum inside long hollow tubes of space and time. These tubes, known as cosmic strings, would still contain the early nothing of the universe. The tubes, in turn, might also get tied into knots or kinks that couldn't come out.

None of these possibilities has been ruled out—or proved, either. But as Michael Turner points out in "The Cosmology of Nothing," "Cosmology offers a unique laboratory for exploring the 'physics of nothing.'" The universe as a whole offers up a whole new realm of possible shapes and forms that nothing at all can assume.

Chapter 6

NOTHING GETS STRUNG OUT

String theory is a piece of twenty-first-century physics that fell by chance into the twentieth century.

—EDWARD WITTEN, Institute for Advanced Study, Princeton.

WE'VE COME A LONG WAY from the bleak emptiness of antiquity. Modern physics animated the void and gave it structure. Today we know the vacuum fluctuates, creating particles and forces. The background of spacetime warps, producing gravitational pull.

Still, it is a relatively simple picture: large-scale spacetime rolls along smoothly while small-scale vibrations of the fields set nothing all atwitter.

Is that all there is to nothing? Is it enough?

The most cursory glimpse at the world we perceive through our senses is far richer than this nothingness can seem to support. Things sprout, drip, spew, dive, wander, branch, breathe, trickle, fizz, pierce, curl, spiral, branch, shine, flicker, and fade. The shapes and motions are rich and irregular. The nothing we have met so far seems far too neat to give rise to all of messy reality.

In truth, of course, it's not all that hard to get complexity from simplicity; it happens all the time. From simple electrons, protons, and neutrons, everything in nature grows—argon and krypton, gorillas and begonias, skyscrapers and stars. Intricate crystals spring from simple seeds, gurgling babies, from single cells; consciousness emerges from nets of virtually identical neurons.

And yet, there would be a more satisfying symmetry to the relationship between something and nothing if the ultimate backdrop for everything were more than sinusoidal vibrations of fields and gently rolling four-dimensional landscapes.

As it turns out, nature has been dropping hints over the past few decades that nothing may indeed be far richer than even Einstein and the quantum mechanics thought. This new geometry of emptiness curls, folds, loops—even rips and tears—in eleven (or perhaps twelve) dimensions of space and time.

The more complex landscape comes with a set of ideas generally known as "string theory"—short for superstring theory—and more recently M-theory, which can stand for magic, mother, mystery, matrix, or membrane theory, almost according to taste. As described by M-theory, the entire universe arises from the harmonics of vibrating strings, membranes, and blobs in eleven dimensions. These unseen dimensions curl around one another in strange, convoluted shapes, forming holes, knots, and handles, leaving pieces of the universe oddly isolated, perhaps stranded on islands or hanging on to the rest of the cosmos by tenuous threads.

Our entire tangible universe may be trapped on a nine-dimensional membrane attached to a larger ten-dimensional universe only by gravity.

String theory strikes some as strangely beautiful, others as nothing short of bizarre. Ignored by all but a few smitten followers until a few years ago, it has suddenly become part of the common parlance—tossed around at physics meetings with as much familiarity (and sometimes disdain) as "Monica" and "Bill." Recently, its success at solving certain long-standing puzzles has won over even some of its most vocal critics.

Physicists first became attracted to string theory because it, and it alone, resolves the glaring mismatch between the laws that rule the large-scale cosmos (Einstein's theory of gravity) and those that run the microcosmos (quantum mechanics). In string theory, not only do the two sets of rules get along together, they need each other; they require each other to exist. On close inspection, even the unruly quantum spacetime "foam" becomes the raucous song of vibrating strings.

The trick to making this difficult marriage work seems deceptively simple: The main reason that quantum mechanics and gravity don't mesh is that the two together produce infinite solutions to equations. Everywhere gravity meets the quantum, infinity pops up, and infinite solutions are nonsense. They are the dead canaries that tell physicists the laws of nature are breaking down.

But the infinities appear in part because present-day theoretical physics allows particles to be infinitely small and space and time to squeeze down to infinitely small specks. Properties of space, time, and particles approach the dreaded zero with impunity. And zero always opens the door to infinity.

String theory solves that most fundamental of problems by doing away with infinitely small particles. It doesn't allow infinitely small anything. The loop of string is the smallest allowable size. Like glasses that blur the view, strings erase the problem. The incompatible partners, gravity and quantum theory, fit snuggly together out of sight, under the covers of size. Indeed, because the strings prevent anything from getting infinitely small, they smear out many previously troubling properties of space and time at infinitely small scales. "You never get to the point where the disasters happen," explains physicist Nathan Seiberg of the Institute for Advanced Study. "String theory prevents it."

And if string theory is right, it would do a great deal more than mend the rift between gravity and the quantum: It could explain almost every outstanding question in fundamental physics—from how the quarks are trapped inside the vacuum to why the vacuum is the way it is in the first place. Because string theory brings together all the laws compatibly under one happy roof, it could well reveal that every tick of a clock, every barking dog, every dying star, flows from a single master principle, a single grand equation.

Fittingly, the power from string theory arises at least in part from the fact that it embodies a more perfect symmetry than general relativity or quantum theory alone; from its higher-dimensional perspective, string theory naturally embraces a multitude of seemingly disparate realities.

NOTHING TO SEE

Why extra dimensions instead of little green people? If they were small enough, we couldn't see them either.

—Brian Greene

What is string theory exactly? It says simply that the building blocks of nature are no longer particles, but vibrating strings of some fundamental, unknown "stuff." Depending on the exact geometry of the vibrating string, it will produce different harmonic chords, just as a piano produces a different sound than a flute.

In turn, the way the string vibrates determines a particle's mass, its electric charge, its spin, and any other properties. By changing its mode of vibration, the string, in Andrew Strominger's words, "can masquerade as an electron or a photon or a graviton or any other particle of nature." In a sense, the resonant patterns are equivalent to the particles, the universe the symphony they sing.

While some strings have free ends, others curl into loops. They interact with one another by joining and splitting off again, branching into new strings and loops, growing and shrinking. If a universe built up from pointlike particles resembles a city made of Lego blocks, the universe of strings looks a lot more like a tree.

We don't see the strings, in part, because they are unimaginably small. Consider that it takes about a million atoms to cross the period at the end of this sentence, and that it takes about 100,000 protons to equal the size of an atom. Then think how many protons it would take to cross New Jersey. That's how many strings could, theoretically, fit inside a single proton. For the more quantitatively minded, the string scale is 10^{-33} centimeters (a fraction with 1 in the numerator and the number 1 followed by 33 zeros in the denominator).

But size is only one reason we can't see the strings. They also curl around dimensions unfamiliar to our senses. Consider an ant living on a beach ball. To the ant, the beach ball seems a flat infinite space—like the earth did to our forebears. Or think of a garden hose: seen at a great distance, it looks like a one-dimensional line. In fact, it curls around a second (from our perspective unseen) dimension.

If string theory is right, we are all ants living on garden hoses that curl and twist into at least six unseen dimensions. Sometimes, dimensions wind around one another, creating extra-energetic (that is, massive) particles. Sometimes, the dimensions flatten out, or deflate, like a three-dimensional beach ball flattened by a truck. These transformations can produce astonishing effects, such as turning something that looks like an elementary particle into something that looks more like a black hole. And vice versa.

The exact shapes of these extra dimensions are critical. Just as the exact shape of a trumpet or violin determines the sounds it can make, so the topology of the extra dimensions determines the possible particles produced by the strings. The topology of the extra dimensions in which the strings vibrate, in other words, determines the ingredients available to make our universe.

Every time you sweep your hand through space, it also sweeps through all the extra curled-up dimensions. But they are so small, and your hand circles through them so fast, that it gets back to its starting point before anyone can notice.

How do we know they're there if we can't see them? Why do so many physicists put so much faith in something so far removed from the palpable, testable, world? The reason is, the theory seems to work. Physicists have trust in string theory for the same reason most people have trust in computers or jet planes and other things they don't understand—these things work for us; they take us where we want to go; and all things considered, they seem to be fairly reliable.

No one knows, as yet, exactly what string theory will turn out to be; but because it has performed so many amazing feats, many physicists are willing to wait to see where it takes them. And at least until experiments prove them wrong, they are willing to trust that something more real than little green people lies hidden in the ultrasmall world of extra dimensions. "Nobody in this field is clever enough to have invented something like this," said Strominger. "It's clearly something that we discovered."

A STRING OF MIRACLES

It's as if some guys had set out to design a better can opener and wound up with an interstellar space ship. And then, they spent ten years looking at this thing and saying: "This won't work as a can opener; it's bigger than the average kitchen."

—SIDNEY COLEMAN

Ever since it dropped into the laptops of physicists almost thirty years ago, string theory has been surprising people with its staying power. It was discovered by accident in an attempt to solve the problem of how quarks stick together inside protons and neutrons. At first, no one even knew that the equations described strings. The theory predicted impossible, faster-than-light particles. It did not include particles of matter, but only particles that transmit forces. It seemed to require twenty-six dimensions.

Better explanations came along for the trapping of quarks, and most of the physics community looked the other way. Still, a few stalwarts stuck with strings, impressed by a string of "mathematical miracles" the theory produced. "It looked a little crazy," remembers Caltech physicist John Schwarz, one of the few loyalists. "But I felt such a beautiful mathematical structure had to lead someplace."

Even seemingly insurmountable obstacles turned out to be opportunities. For example, one early version of string theory produced a strange particle that didn't seem to fit anywhere in the standard picture of the physical world. "Eventually, we decided to stop trying to get rid of the thing and take it seriously," Schwarz said. In a classic case of looking at what everyone else had seen, but thinking what no one else had thought, Schwarz recognized the problem particle as a graviton, a "particle" of gravity. String theory went from a theory of quarks to a "theory of everything."

Still, until five years ago, most physicists dismissed the theory as so much mathematical navel gazing, as untestable as counting angels on the head of a pin. But in a remarkable turnaround, since 1995, string theory has come out of the closet. No longer on the fringe, it has become respectable, perhaps—in some form—inevitable. "It really migrated to center stage," said Greene, whose book on string theory, *The Elegant Universe,* was soon on the best-seller lists. "People are convinced that it's not a fad; it makes a believable—if tentative—claim to being the final theory [of everything]."

What changed is that string theory solved some long-standing problems previously out of reach of any other theory.

For example, string theory made sense of a proposal made by

Stephen Hawking more than twenty years ago that black holes hid a precisely calculable amount of disorder inside their borders. It was hard for physicists to imagine how something as featureless as a black hole could be anything but highly ordered. A room with one piece of furniture is always more orderly than a room with one hundred pieces of furniture. The more pieces to a puzzle, the bigger the mess it will make when it falls on the floor. And a black hole, for all its bizarre properties, is a very elementary object—a simple warp in spacetime, completely described in a few simple terms.

In 1996, Strominger and colleagues showed how black holes could be constructed, in a sense, from strings. What's more, their calculations produced the precise amount of disorder Hawking predicted should be found. Indeed, along with Greene and others, Strominger showed that black holes could transform into elementary particles, like water freezing into ice, under the right conditions.

The result propelled string theory to the forefront. "It's not just that it's mathematically beautiful," explains string theorist David Gross. "It's physically beautiful. If you have a theory that can come to grips with a problem that's been around for twenty years, it builds confidence."

In addition, string theory vastly simplified the mathematical tools for dealing with more traditional problems in four-dimensional space, including standard particle physics. These results, said Seiberg, were things other physicists "could relate to. It solved problems they'd been bothered by."

Together, the black-hole results and the Seiberg-Witten equations (after Nathan Seiberg and Edward Witten, who developed them) forced even confirmed skeptics to take a second look. "These were really attractive results," said Coleman. "[String theory] was no longer just the only game in town. It was paying off big."

There are even some very preliminary indications that string theory may have an answer to the question of why elementary particles appear to be grouped into three distinct families, each falling into an entirely different range of masses, rather like molehills, mammoths, and mountains. It seems that the number of particle families is tied directly to the number of holes in the curled-up extra dimensions. "This is the kind of result," says Greene, "that makes a physicist's heart skip a beat."

And recently string theory even spawned something that looks potentially like the mysterious antigravitational-like force that appears to be pushing the galaxies apart. If so, it could help resolve serious paradoxes about the energy of empty space that have plagued physics since Einstein.

"It could be a coincidence," said physicist Joseph Lykken of Fermilab. "But it could be something interesting."

THE ORIGAMI UNIVERSE

In this space, we don't know where we are; we don't know what the topology is. Pieces may be connected or disconnected by little filaments or islands. There is a possibility that . . . we are not seeing the bulk of space-time.

—Fermilab physicist JOSEPH LYKKEN, Cosmo-98

Originally, string theory seemed to work best with ten dimensions: one time dimension, three "normal" space dimensions, and six curled-up "garden hose" dimensions. The much-heralded results mentioned in the previous section, however, developed from an expansion of strings into another, eleventh, dimension of space. In M-theory, the fundamental one-dimensional strings grow into two-dimensional membranes—sheets, like bubble walls, that merge and meld in various ways to create everything in nature. In fact, the extra dimension gives rise to a whole zoo of brany new objects, including three-dimensional blobs (known as three-branes), nine-dimensional sheets (known as nine-branes), and just about everything in between.

A zero-dimensional point is not, strange to say, a no-braner. In this stringy nomenclature, the number attached to the brane is the number of spatial dimensions. However, time also has to be factored in. Even a zero-brane still has one dimension of time; that is, it's a point in space that hangs around for a while. Something with zero dimensions of space or time, however—like the instanton—is a minus-one-brane, a mere instant in both space *and* time. As Stanford physicist Lenny Susskind put it recently: "A zero-brane is forever; a minus-one-brane is here today, gone tomorrow."

Under certain conditions, branes can deform into one another, tearing spacetime to create new submicroscopic landscapes. Going back to the garden hose analogy, this means that the walls of the hose itself could expand into extra dimensions or shrink the hose into a string that tied itself in knots. The hose could sprout holes, or it could repair holes already present by wrapping them up in sheets. A two-dimensional blob could intersect another blob of three or more dimensions; these blobs could fold around each other, adding new dimensions and shedding old ones, blooming into a wild landscape of multidimensional forms.

What used to be a universe of simple vibrating fields and warped spacetime, in other words, has transformed into a universe of undulating—almost organic—origami.

While membranes may seem to complicate the landscape of strings, the truth is that this extra, eleventh, dimension (attributed mostly to the work of Edward Witten) simplifies matters enormously. Before M-theory, there were at least five different string theories with different configurations of dimensions, and each seemed to have little in common with the others. M-theory showed that all five string theories were part of a grander scheme. Unbeknownst to everyone, they were all aspects of one another.

"There is now one string theory," Lykken announced.

M-theory tied together string theory in a way that gave it the credibility it needed. The situation is strikingly similar to the state of physics before the discovery of quantum theory, when light seemed to be either waves or particles. Later, it was shown that light—like matter—has *both* wave and particle aspects. Both are different aspects of the same entity, just as heads and tails are two sides of a coin.

This wave/particle duality is mirrored in several important dualities recently discovered in string theory. "It means you can look at one physical property in two very different ways," explains Greene. To wit: There are string theories that work without any reference to spacetime that seem equivalent to theories that require spacetime. There are string theories in which as things get smaller, they also get bigger. There are string theories that seem to be something like holograms of others: like the ephemeral three-dimensional "object" that seems magically to emerge from the two-dimensional surface of your credit card, some string theories create a higher-dimensional universe using only the information on a lower-dimensional "surface." It's as if you can obtain all the information about everything inside a room merely by understanding the physics of the walls. (More on this later.)

All these dualities suggest that string theorists have been looking at the same animal, only some have discovered the tail, while others have found the ears or glimpsed a snout. The problem is, they still don't know what kind of animal they're dealing with.

HONEY, I BLEW UP THE UNIVERSE

It's an astonishing fact.
But you cannot rule out that possibility.
—ANDREW STROMINGER

Intriguingly, developments over the past few months suggest that some of the new dimensions described in string theory may actually be within reach of experimentalists, if not today, in ten or twenty years. This would be very good news for string theorists. Skeptics have found it relatively easy to blow off strings as so many rings of theoretical smoke. After all, the vibrating strings and unseen dimensions that hold them were too tiny ever to be seen in experiments. And a theory that can't be tested is about as relevant to a physicist as a bicycle is to a fish.

But what if the unseen dimensions were much larger than previously thought? Big enough to see in the relatively simple experiments now in place or in those planned in the next dozen years? In fact, the idea is rapidly gaining respect, if not outright adherents, among many scientists.

If it's true, it would mean that "string theory is just out of reach of experiment," according to Lykken. Even more important, it would mean a whole new way of solving a host of thus-far elusive problems—mysteries ranging from the unexplainable weakness of gravity to the unaccountable existence of matter in the universe at all.

The key to sensing the extra dimensions would be the omnivorous appetite of gravity. We could never "see" the extra dimensions with our eyes, because electromagnetic radiation doesn't travel out of our familiar four-dimensional world of height, width, breadth, and time (if it did, we would have noticed something missing). We can't touch the extra dimensions, because touch is a function of molecules, and molecules operate on chemistry—also electromagnetic in nature. Nuclear forces—or even the properties of the vacuum itself—are similarly confined to our four-dimensional universe.

But gravity is a different story. Gravity goes everywhere. Because gravity *is* the warp of spacetime, it goes wherever spacetime goes—including into extra dimensions. "Nothing can hide from gravity," says Strominger. " Gravity knows about all forms of matter and energy." And gravity—because it's so weak compared to the other forces—has not been well tested much below the scale of a grain of rice. Perhaps it *can* communicate with extra dimensions and on a much larger scale than anyone previously imagined.

According to this new scenario, the everyday universe we live in is trapped on a thin membrane something like the world inhabited by characters in movies who play out their lives confined to a screen. Unbeknownst to these shallow two-dimensional players, a larger universe spreads out into as many as six extra dimensions, like theaters in a multiplex. And while we are stuck as firmly in our membrane as Rhett and Scarlett are stuck on the silver screen, gravity can escape, leaving behind experimentally detectable tracks.

If gravity leaks off the membrane, that could explain the puzzling disparity between the strength of gravity and all the other forces. The disparity would make perfect sense, according to Stanford physicist Savas Dimopoulos and his colleagues, if gravity alone could ooze off our membrane into the larger universe, while the other forces were stuck here at "home."

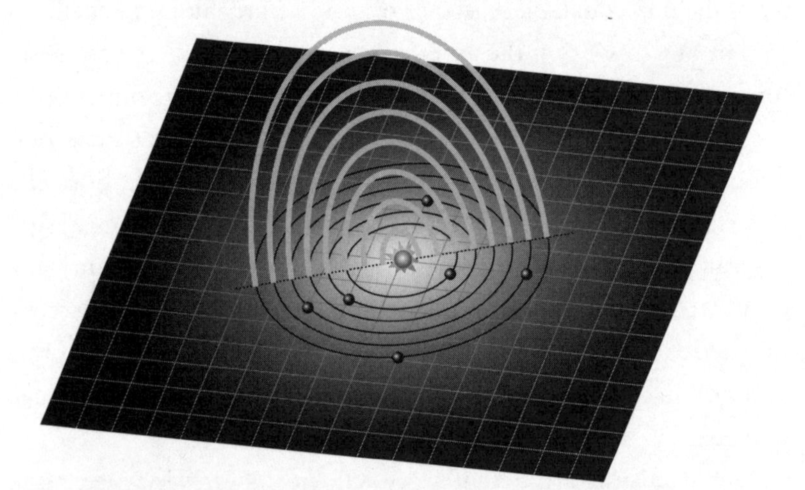

The light from the Sun—along with nuclear forces, chemistry, and everything else—resides in the familiar grid of four-dimensional spacetime. Does gravity also "leak" into higher, unseen, dimensions?

Imagine our three-dimensional universe as the scum that forms on the surface of a pond. All the forces that make up our everyday universe—electricity, magnetism, and nuclear forces—would be trapped inside this surface. According to string theory, they would be vibrations, or waves, that could vibrate in this surface. However, another, deeper dimension would lie beneath the surface, like the water underneath the scum. Only gravity could make waves in this deeper dimension. "Our universe is sort of stuck at the edge of the (deeper) dimension," explains Strominger.

In fact, most of gravity would "live," as the physicists like to say, in the bulk of the water. Only a tiny portion of the total gravitational force would be felt on the surface. "The reason why gravity is weak is that it lives far away from us," says Dimopoulos. "In a way, it's a very simple idea. Faraway things don't interact as much as nearby things. You don't have to worry about a faraway lion as much as a nearby lion."

These extra dimensions that gravity seeps into cannot, according to most scenarios, be just any size. If they extended as far as the Sun, for example, the leaking of gravity into the extra dimensions would have weakened the pull of the Sun on the earth. By now, astronomers would have detected any variation from Newton's (and Einstein's) well-tested laws. In fact, gravity has been well tested down to about the size of a millimeter. That's why any extra dimensions are generally thought to be flea-sized or smaller.

But below a millimeter is anybody's guess. In Dimopolous's scenario, two extra dimensions curl into cylinders like straws with a diameter of about a millimeter. If we could probe gravity with sensitive enough experiments, we might be able to sense their presence. For example, Newton's laws might change as gravity "leaks" into the extra dimensions. Specifically, gravity would no longer fall off as two objects move away from each other according to Newton's familiar $1/r^2$, where r is the distance between the two objects. Instead, it might fall off faster. In Dimopoulos's scenario, with two extra dimensions, gravity would fall off by $1/r^4$. With two extra dimensions to leak into, gravity would weaken with distance much faster.

In less-radical versions of this idea, extra dimensions would be much smaller than a millimeter, but still much larger than the scale of tiny strings. If the dimensions are, say, the size of an atomic nucleus, the Large Hadron Collider now under construction in Geneva could conceivably detect them within the next decade. Energy leaking out of our membrane might show up as missing energy in particle collisions. Or the extra dimensions would make themselves felt as entire new families of particles. That is, every familiar particle would also have an additional way to vibrate in the extra dimensions. That would give it an extra source of energy, and therefore mass.

"We'd begin to see a whole new kind of matter," said Gross. "That would make string theory a little less metaphysical." If the physicists get very, very lucky, the first signs of higher dimensions could pop up even sooner—perhaps within the next few years—at Fermilab near Chicago.

And there are still other possibilities. Along with colleagues, Lisa Randall of MIT and Princeton has been exploring the possibility that gravity changes strength dramatically in various parts of this higher-dimensional world: we just happen to live on a slice of it where gravity is weak. Randall's approach allows the extra dimensions to spread out in space like the everyday dimensions of height, width, and depth; we could not sense them because there's no way—other than gravity—to communicate with them. But they would not have to be small or rolled up; they could be infinite in extent.

Just like people, branes could come in a wide range of types and sizes: curled up like straws, spreading out infinitely into space, or simply folded up. Dimopoulos has recently suggested that the universe we live on might be part of a huge megabrane that is folded up like an accordion into as many as several hundred parallel surfaces. We would live on one of the surfaces; the others could be filled with stars and—who knows?—planets. We would feel their gravity. But we would have no other way to know anything was out there because only gravity could leak directly from one surface of the folded "accordion" to the other. Light would have to travel the long way around, taking all the corners. By the time it got to us, no one would be left in our universe to know or care.

NOTHING VENTURED

You have this whole new setup now. There are some hard problems out there that we haven't been able to get at. Maybe there's something lurking here which will help us solve some of these problems.

—physicist LISA RANDALL

Up until recently, physicists tended to look to our own familiar universe for solutions to fundamental problems. And why not? That was, after all, the only universe they knew. But the possibility of detectable extra dimensions has changed everything. There might be more out there than anyone dared to guess: more nothing as well as more something. Either one could make itself felt in our little slice of universe.

Take, for example, the notorious case of the missing matter. Somewhere out there, something is generating a strong gravitational pull (or warping spacetime, as you wish) in ways that are obvious and visible: stars and galaxies and the expansion of the universe as a whole follow paths in spacetime that are clearly warped. But visible matter cannot account for this behavior.

Needless to say, extra dimensions are an obvious place to hide the dark matter. Since gravity spreads into all dimensions, it could easily yank on our thin bubble membrane from its home elsewhere in the bulk of the universe. Signals from outside could deform our membrane (warp it, if you will) just as a gust of wind from outside a bubble could deform the soap film. This would be a scientifically unsatisfying solution to the dark matter mystery, however. If the dark matter can only interact with us through gravity, Dimopoulos points out, then there is no other way to detect it. So there would be no way to confirm its presence or properties. In a sense, its "discovery" would amount to a tautology.

But explaining dark matter and other gravitational anomalies is only the tip of this multidimensional iceberg. Once you introduce the idea of membranes sitting in a larger space, it's only natural to wonder what else is going on out there. After all, we learned a long time ago—starting with Copernicus—that there is nothing unique or special about Earth's place in the solar system, or the Sun's place in the galaxy, or our galaxy's place in the cosmos. Why should our membrane be any more special than our star? "It would be too anti-Copernican to assume that the whole purpose of [the space beyond our membrane] is our existence," said Dimopoulos. "The most natural thing is that there are other branes."

It's certainly possible (maybe even probable) that the laws of physics on these other membranes are different from ours. It's possible that forces other than gravity—forces we don't yet know about—leak off these other membranes and somehow make it into our four-dimensional world. If so, they could have far-reaching effects, and perhaps even explain some of physics' most difficult puzzles. For example, these forces might account for many strange quantities in physics that are almost zero, but not quite: The mass of the neutrino appears to be practically nothing . . . but not quite. The amount of matter and antimatter produced in the early universe appears to be nearly identical . . . but not quite. Theories suggest that the proton is stable almost forever . . . but not quite.

What makes these numbers so terribly, tantalizingly small? Perhaps these phenomena are all caused by fields living very, very far away from our own membrane. Dimopoulos even speculates that the wide range in masses of particles might be explained by the proximity of their "home" membranes to ours. The electron weighs less than its otherwise identical twin, the muon, in other words, only because the "field" that determines the mass of the electron originates farther away from us than the field that determines the mass of the muon.

In this universe, everything depends on distance, that is to say, geometry. Change your position in the multidimensional universe, and everything else changes as well. Below the scale of rice, gravity could change beyond recognition. And far bigger surprises could be in store. Dimopoulos's latest work predicts that previously unknown forces reaching us from membranes far beyond could be a million times stronger than gravity.

In fact, Dimopoulos has gone so far as to suggest—only partly tongue in cheek—that people could store information in these extra dimensions. "I'm talking about making a profit, telling Bill Gates that there is money out there," he said. "This is a possibility that hasn't been investigated. But there could be technology there."

Not many string theorists expect the experimental search for "large" extra dimensions to pan out. However, the fact that they are theoretically possible is important. The new ideas show that writing off string theory as forever untestable is "ludicrous," says Strominger. "This kind of structure never occurred to anybody before, but it turns out it's very natural." If nothing else, he explains, "it shows how limited our imaginations have been up until now. It shows how little we actually know about the universe around us just beyond the limits of what we've actually measured."

STRINGS OF WHAT?

You're not allowed to ask that question.
That's it. It's just the thing.
—ANDREW STROMINGER

It's nothing. It's everything.
—DAVID GROSS

Institute of Theoretical Physics director David Gross doesn't like to call the theory-formerly-known-as-string-theory by any of its current names, including M-theory. He thinks that "U theory" is far more appropriate. The "U" stands for unknown. "There is a theory," he says by way of explanation, "but we don't know what it is."

To be sure, physicists have faith in string theory because, in some important senses, at least, it seems to work—just as people fly in airplanes because they work. The difference is, *somebody* understands how jets work. And *no one,* as yet, understands what underlies string theory.

Brian Greene compares string theory to a computer dropped into the nineteenth century. People could play with the computer and perhaps even get it to do seemingly miraculous things. But no one could, even in principle, figure out how the computer worked for the simple reason that the essential science behind modern computers—quantum mechanics—hadn't yet been invented. And until string theorists get some sense of the basic science behind their wonderful new toy, they won't fully be able to understand it, either. Greene describes the modern-day string theorist as an excited but puzzled child who receives a wonderful gift on Christmas morning and can't figure out how it works because the instructions are missing. "Today's physicists are in possession of what may well be the Holy Grail of modern science," he says, "but they can't unleash its full predictive power until they succeed in writing the full instruction manual."

In many ways, string theory was developed backward compared to the normal process of discovery. Usually, scientists discover some broad overarching view of nature, and later they fill in the details. In string theory, they have the details, but not the broad, contextual, view. What they have, in Strominger's words, is the "most fantastic set of interconnected rules ever known. Different aspects of the theory mesh into each other in a very intricate way." But no one knows exactly what the rules describe.

So what do string theorists need to know about their theory—

besides its proper name? Some very basic things. For openers, what is the ground state of the theory—the bottom line? In other words, what is the vacuum in which strings live? Normally, any theory has a single fundamental lowest-energy state, or vacuum state, from which everything else springs. In a way, the vacuum is the zero point of the theory, the starting point for all that follows. Like a ball that's rolled downhill and has landed in a gutter, like the pendulum of an old-time clock that's stopped swinging and hangs there still as stone, the ground state is where everything stops.

But string theory has no single vacuum state. "It has zillions," says Gross. And that is an unprecedented—and highly unnerving—situation.

STRANDED IN NOTHING

The real world is not a member of any of these {ground} states that we know something about.

—LENNY SUSSKIND

Think of a ground state as the bottom of a bowl. Throw a marble into the bowl and it comes to rest there. There's nothing at all going on at the bottom of the bowl, and that's how physicists like theories—and their ground states—to behave.

The ground states in string theory, however, look more like continents adrift. The odd configurations of many dimensions don't create anything as tidy as a simple bowl; instead, they create weird geometrical forms and each configuration is its own family of vacuum states. Imagine, instead of a bowl, a collection of exotic landforms: continents, islands, and peninsulas, with rivers, lakes, and mountains carving the landscape into strange topologies. Each of these landforms is a family of vacuum states. There are tens of thousands of such families. And each family contains an infinite spectrum of individual vacua.

This presents more than a theoretical dilemma. The fluctuations around the zero point, or ground state, correspond to the particles that any theory should produce. If you don't know the ground state, you don't know what particles to tell experimentalists to look for. So string theory lacks the most essential property of any believable physical theory: the power to predict what will be found in experiments to prove the theory is right. Right now, there are simply too many vacua in string theory to even begin to predict what they might produce.

Worse yet, not one of the vacuum states discovered so far looks anything like the one our universe inhabits. If the families of ground states are continents adrift, our vacuum is a small island far removed from everything else—or at least far removed from everything so far discovered. We may be attached to the larger universe by a hidden spit of sand, buried under who knows what. In fact, it appears that all possible vacua in string theory are somehow connected, if tenuously, to one another. "In principle, if you're in any one of them, you could access the others," explained Susskind.

Finding the right vacuum among a vast number of possible solutions will be much harder than finding a needle in a haystack. It will be more like finding a small island while blindfolded in the midst of a raging storm.

THE HOLODECK UNIVERSE

It's not only a better description {of the physical world}. It's the only description. . . . I'm absolutely convinced the holographic idea is right.

—LENNY SUSSKIND

Strings theorists don't even know, exactly, the right number of dimensions for string theory. That's partly because in some cases, different string theories are related to each other like the two-dimensional engraving on a credit card is related to the three-dimensional image that floats above (or below). So a string theory with eleven dimensions of space and time might be equivalent to another theory with only five. "It's all somehow dependent on how you look at it," said Gross. Obviously, there's a bigger picture here that incorporates all the different dimensional views. For now, at least, no one knows what it is. "Nobody even has a proposal," said Gross.

How could an eleven-dimensional universe also be five-dimensional? It would happen in the same way that a three-dimensional image gets constructed in space from the information encoded on a two-dimensional surface. The hologram on a credit card, for example, appears to pop out like a solid object even though your fingers confirm that the plastic card is flat. The way it works, roughly, is simple: Just as the grooves on a phonograph record encode all the information needed to reproduce a sound, so all the information needed to produce the image in a hologram is encoded on a two-dimensional surface. When you put a needle on the record and hook it up to speakers, the original sound is reproduced. When you shine a light on or through a holographic plate, a three-dimensional image is produced.

In Susskind's view, the universe is a hologram that is created from information encoded entirely on a lower-dimensional boundary.

At first glance, the notion seems utterly impossible. After all, intuition tells us that a three-dimensional universe contains more information than a two-dimensional boundary around it. So how could you create the former from the latter?

Imagine, for simplicity's sake, that the universe consists entirely of one kind of fundamental building block—say, a sugar cube. Say the universe is a large cube made out of smaller sugar cubes. A cube with the dimensions 10 × 10 × 10 sugar-cube units would produce a universe containing 1,000 fundamental bits. A single face of the cube, however, would only contain 10 × 10—or 100 bits. How could you possibly encode all the information necessary to produce the 1,000-bit universe on a 100-bit surface?

Astonishingly, it turns out you can. A two-dimensional surface can contain all the information needed to reproduce a three-dimensional solid, or "bulk," as the physicists call it. No one was more surprised by the result than Susskind himself. "This was something that was totally unexpected," he said.

Laypeople shouldn't worry if they can't fully understand how this works. Even the physics community, Susskind admits, "is having trouble getting its head around this. It's extremely puzzling and mysterious." However, the mathematics seems extremely solid. Worked out in detail by physicist Juan Maldecena of Harvard University several years ago, it's now so convincing, said Susskind, that "most of us think it's an absolute fact."

It would mean, in essence, that if you want to describe the universe and everything in it—gravity, particles, planets, books and their readers—the correct information you need to do that lies not on the three-dimensional "interior" we take to be the real physical world. Instead, it is encoded on a lower-dimensional surface. Like a hologram, the universe is a projection of the information on this boundary into space.

One reason some physicists like this idea is that reading the laws of nature on the boundary is a lot easier than it is in the messy interior. As Maldecena showed, the equations on the surface make far more sense. To go back to the sugar-cube analogy, if you tried to figure out the rules of the sugar-cube universe by looking at each of the 1,000 cubes, you would get lost in a labyrinth of information. "It's too confusing in there," says Susskind. "There are too many possibilities. There are too many ways things can wiggle. By the time you add up all the wiggles, everything gets out of control."

The equations as they appear on the boundary, on the other hand, seem to embody just the kind of sensible, consistent laws physicists have been seeking for decades.

Of course, there's much more work to do. No one knows yet precisely how to decipher the code written on the boundary and project it to create the universe we experience and measure. "We are far from knowing in detail how to translate back and forth between the surface to the bulk," says Susskind.

Even more critical, the holographic idea works so far only in extremely, bizarrely, curved spacetime, like that surrounding black holes. In fact, Susskind and his colleague Gerard 't Hooft first stumbled on the holographic principle while exploring what happened to the information lost down the deep throats of black holes. They discovered that information about everything that goes into a black hole is, in fact, stored on its boundary—the event horizon. By extension, they found that all the information that goes into, say, a room can be stored on its boundaries, or walls. If you try to put more information into a room than can be represented on its walls, the room turns into a black hole that is bigger than the room.

THE DOOM OF SPACE AND TIME

We need to find the dictionary that defines space and time in this new language.

—GARY HOROWITZ, University of California at Santa Barbara

Most of all, physicists don't know what happens to space and time in a universe of strings. For one thing, the very fact that a string theory with no space and time can be equivalent to another theory with space and time suggests that space and time aren't really fundamental. For another, space and time behave bizarrely in the universe of strings. "Space and time get confused," as Seiberg puts it.

For example, the algebra used to describe space and time appears to undergo some radical transformations on the scale of strings. Normally it doesn't matter whether you multiply x times y, or y times x; 5×3 is the same as 3×5. It doesn't matter which number goes first. But at the string scale, it does matter, and 5×3 does not necessarily equal 3×5. Quantities that are normally interchangeable suddenly are not. It normally makes no difference whether you salt your eggs before drowning them in ketchup. But it does make a difference whether you scramble the eggs before you cook them. The outcome changes when you reverse the order. The same appears to happen in the space (and possibly even time) of strings.

There are stranger things, as well. As distances get smaller and smaller in string theory, things don't collapse into points. They can't, because the string is the smallest thing there is. Instead, the more you squeeze the string to make it smaller, the bigger it gets. "This is an amazing fact," says Dimopoulos. "If you try to bring two strings very close together, you put so much energy into it that you end up creating a very long spaghetti out of the energy." Try to pin down a string, in other words, and instead you create string spaghetti. In a sense, squeezing the string turns it into a particle accelerator, because every vibration of the newly created string is a new particle.

Ultimately, all this confusion is a good thing, says Seiberg, because "it's telling us that the traditional understanding of space and time will evaporate and there will be a more interesting and subtle result." However, all known physics—including string theory—still operates upon the well-trod stage of space and time. Somehow, ordinary space and time need to emerge from the spaceless, timeless realm of strings. Somehow, the four-dimensional fabric that warps to keep the planets in place around the Sun has to be created; somehow, there has to be a background for the constantly wiggling fields. Somehow, there has to evolve the space and time in which string theorists themselves exist and do their work.

So far, there's no clear way to weave the strings into the familiar backdrop our universe calls home. That doesn't mean the road won't be found. After all, "wetness" isn't a property you would expect from studying the equations for water molecules, and the richness of "color" wouldn't be evident from studying atomic quantum states alone. And yet—somehow—wetness, color, time, temperature, and taste all emerge from unfamiliar properties of the quantum realm. Something similar could be behind the relationship between time, space, and strings.

Greene suggests that individual strings may be shards of space and time that only begin to resemble everyday space and time when they vibrate in resonance. But the real physical meaning of this suggestion has yet to be worked out. "It's a huge challenge which we have not lived up to," said Strominger. "String theory has been giving us a lot of clues, but we haven't been able to put them together into a unified picture."

The specter that string theory dissolves space and time is unsettling, to say the least.

"When we talk about space and time, we think there is something there, and we live in it," said Gross. If there's no space and time, "that's very disturbing. Where are we? When are we?"

While the breakdown of space seems (at least to some) ultimately plausible, the breakdown of time seems (to almost everyone) inconceivable. And yet, Einstein taught us that space and time are equivalent, so time's time has surely come.

As Gross posed the question to his colleagues at a recent meeting, "What *about* time? Time is of the essence."

The almost unfathomable scenario of a universe without time in turn calls into question the very notion of causality. If time can break down, how can one event be placed clearly "before" or "after" another? Hypothetically, if there is no clear difference between now and the instant after, how can we say whether the gunshot caused death—or death caused the gunshot? "We normally think of causality as a basic property," said Horowitz. "Something affects something else. But when you're getting rid of space and time... are we sure that causality is going to be preserved?"

New views of time could lead to even more bizarre consequences, for instance, two or more dimensions of time, a theory being worked on by USC physicist Itzhak Bars, among others.

LOOPY?

The biggest problem faced by the background-dependent approaches is in getting rid of the background, while the biggest problem faced by the background-independent approaches is restoring the background.
—Lee Smolin

If string theory can't yet create spacetime from strings, there is another theory in the wings that begins with loops (if not strings) and winds up creating something very much like spacetime. While this "loop space" hasn't produced any of the "miracles" attributed to string theory, it has garnered some serious interest. Ultimately, it may turn out that loop space and string theory are intimately related.

Loop space begins with pure geometry: spacetime is a weave made of quantum-scale knots or links, rather like a woven carpet. These are not knots of anything. They are, as Strominger might have put it, "just the thing." The knots are reminiscent of Kelvin's nineteenth-century notion that knots in the ether gave rise to all the particles in nature. Except that these knots give rise to spacetime as well. Spacetime is what you get when you weave together all the quantum-scale loops.

The difference between strings and loops is primarily the starting point of the two theories. String theory emerged from particle physics; it takes place on a background of normal spacetime. It has yet to figure out how to create spacetime from string theory. Loop space, on the other hand, begins with loops of pure geometry; the loops are spacetime; the challenge is to create the rest of the universe from that backdrop.

As one of the founders and main proponents of the theory, Smolin's aim is to find the relationship between loop space and M-theory—a bridge that will link the two. He thinks it's possible that loop space has found precisely what M-theory needs—the missing link, so to speak, that creates spacetime from strings. "A loop in an enormous and complex network could turn out to be a microscopic close-up of the same phenomena that string theory describes as a string moving in a smooth spacetime geometry," he writes. "Or it could not, only the future will tell."

One of the appeals of loop space is that it is, in the end, only about relationships. The universe is nothing but relationships, nothing but geometry. There is no there there. There is only the relationship between one thing and another.

As Smolin puts the question: "Does the world consist of a large number of independently autonomous atoms, the properties of each owing nothing to the others? Or, instead, is the world a vast, interconnected system of relations, in which even the properties of a single elementary particle or the identity of a point in space requires and reflects the whole rest of the universe?"

His answer is that space, time, and matter are all part of a larger web: "There are relationships, and nothing else."

Connections, in other words, may well be all there is to something. The only difference between something and nothing may be that something has connections. Something *is* connections.

As E. M. Forster put it: "Only connect."

Connecting may well be what the universe does best—even exclusively.

WHY US? WHY NOW?

String theory might allow us—indeed might require us—to understand the principle that fixes the initial conditions of the universe, and thus predict its history and fate.

—DAVID GROSS, String Theory at the Millennium conference, California Institute of Technology

A well-understood theory should not only be able to find the right vacuum and predict the particles it can produce; the theory should also be able to explain why we happen to exist in *this* vacuum and not some other. Why is the history of our universe written on this particular blank sheet of paper, and not some other? Why this spacetime? Why, in other words, this particular nothing that produces everything we've come to know?

"What we lack is a principle that chooses that particular vacuum state," Gross explained.

Certainly, a fully functional "theory of everything" should be able to explain how our universe of four dimensions evolved from the eleven-dimensional world of branes and strings. Until string theory can tell us how we got into this vacuum in the first place, it won't be able to explain why the big bang banged or what kinds of conditions led to it and everything that followed.

Most string theorists believe that all eleven dimensions (or whatever the proper number of dimensions is) started as the same size. But for some reason, four (or more) expanded into infinity, while the others curled up. The question is Why? And how? Do all the dimensions try to expand, but some get tangled like knotted shoelaces, wrapped up in themselves or other strings so that they're permanently relegated to a future of lilliputian proportions while their brethren break free? Extra dimensions can curl up in tens of thousands of different ways, producing tens of thousands of different laws of nature and different universes. Why do we live in the one we do? Is it merely because the kind of universe we live in is the only kind of universe intelligent life can inhabit? If the laws of nature were different, would the earth fall into the Sun, would stars go out and molecules never form? Is asking about the ultimate geometry of strings tantamount to asking why anything exists at all? At least some physicists think so. "It's another way of asking, Why is there something instead of nothing?" said Susskind.

The answer to that ultimate "why" may not come out of strings or even fundamental physics at all. It may well be, Gross suggests, that string theory naturally produces zillions of vacuum states, and the one we happen to live in is not determined by any overarching theory, but is simply an accident of history. It just happened. The answer will be found not in fundamental physics, but in cosmology. "We may not find a vacuum state," he said. "We may find a history of the universe."

Chapter 7

NOTHING BECOMES EVERYTHING

When you read or hear anything about
the birth of the universe, someone is making it up. . . .
Only God knows what happened at the Very Beginning
(and so far She hasn't let on).
—LEON LEDERMAN, *The God Particle*

IT ALL STARTED as a joke: The Big Bang. One of those baby names you never quite get rid of, like Wizzie or Sport, no matter how hard you try to get people to call you Elizabeth or Mark. The name just sticks.

Of course, it's not easy to get serious about the notion that the whole universe just suddenly exploded into existence out of nowhere—a hot blast from some 13-billion-year-old past. To all appearances, the night sky is both cold and permanent. The silence alone speaks of eternal serenity; no creaking of gears or scraping of moving parts; only endless black emptiness anchored in place with shiny studs of stars. The universe, common sense tells us, was always here, always will be. The universe doesn't need a reason, or a story. It simply is. Or so everyone—including Einstein—simply took for granted. Einstein felt so strongly that the universe was forever, in fact, that for several years he refused to believe the clear denial of this neat scenario written into his own equations.

But then, suddenly, there it was: the writing on the wall of the universe, plain as graffiti. In 1929, Edwin Hubble looked into space from the top of Mount Wilson near L.A. and saw the signs in starlight that galaxies were racing away from one another.* The universe was expanding like a big balloon. This meant, of course, that if you run the movie backward, the whole shebang must have been compressed together long ago—everything in the universe, starting from a tiny speck in space and time. Hard to imagine, much less believe. Still, the calculations were straightforward. It all made sense. At some point in the distant past, there was good reason to believe, the whole darn thing simply went *Kaboom!* And here we are to tell the story.

Needless to say, "reason to believe" doesn't suffice for science. There has to be confirming evidence. Remarkably, the evidence came raining down on an obscure telescope in New Jersey in the 1960s, when a pair of unsuspecting radio astronomers picked up faint signals that turned out to be the very afterglow of the big bang itself. As awesome to astronomers as a full set of T. Rex teeth to paleontologists, these fossils of creation are tangible imprints from the earliest moments of time.

From that time on, the universe has had a history. And the "origin event" has a name it will never outgrow. Today, cosmologists can dial back the clock to within a few picoseconds of the origin of everything; they can tell you what was in that speck and what it was doing—the players as well as the play. Their theories work so well, in fact, that they have been able to predict—with surprising accuracy—the evolution of the universe, from embryonic particles

*This account is, needless to say, drastically condensed; however, excellent histories appear in many popular science books, including Alan Guth's *The Inflationary Universe: The Quest for a New Theory of Cosmic Origins* (Reading, Mass.: Addison-Wesley, 1997).

to wildly careening baby atoms, to mature molecules, galaxies, stars, planets, and life (not necessarily in that order). It is this close-to-incredible match between what the theories tell and what the universe displays that leaves cosmologists in awe. Forget star-studded nights. (Most cosmologists never look up.) It's the power of the ideas that strikes them dumb.

Remarkably, the entire turnaround in thought from static universe to expanding firecracker took less than fifty years. What's more, it was completely unexpected. That, in itself, contains an important lesson. "We cannot understand the universe by pure thought," Caltech astrophysicist Andrew Lange reminds us. "Nature is more interesting than we can imagine."

THE ULTIMATE FREE LUNCH

Our universe is simply one of those things
which happen from time to time.
—Edward Tryon

It is said that there's no such thing as a free lunch.
But the universe is the ultimate free lunch.
—Alan Guth

That was more or less the whole story up until about thirty years ago. Thirty years ago, asking what caused the big bang—much less what banged, or why it banged, or what happened before the bang—was idle philosophizing or science fiction. Something exploded. Not from anything, or into everything, because it *was* everything. You couldn't really talk about where or when the big bang happened, because it happened everywhere for all time. It *was* all space and all time. And that was all there was to say.

Then in 1973, physicist Edward Tryon, then at Hunter College, proposed that a sufficiently large tweak in the quantum vacuum could produce an entire universe. This is not so outrageous as it seems. It was already old hat (as discussed in Chapter 4) that particles and antiparticles pop in and out of the vacuum continuously and that, given a sufficient jolt of energy, such particle pairs can become "real." Theoretically, at least, anything given enough energy can pop in and out of the vacuum, assuming it appears alongside its opposite number. A particle and an antiparticle; popcorn and antipopcorn; people and antipeople—a suitable pair of anything could appear at least ever so briefly as long as the net amount of whatever popped into existence was zero.

In truth, popcorn and people are far too complex to spontaneously assemble from nothing, but this is not so for subatomic particles—not even in vast numbers. Tryon proposed that purely by chance a huge fluctuation swayed the equilibrium of the vacuum, and out popped everything. Because the tweak was so big, it could only stick around for an instant. But once it started to "bulge," it managed to hijack energy from the expanding space itself (more on that later). "The possibility presents itself," he wrote, "that the birth of our universe was simply the result of an enormous fluctuation . . . at some time and place within a pre-existing quantum vacuum."

The key to making Tryon's idea work is that everything in the universe adds up to nothing. Even Tryon didn't say that everything amounted to nothing—only to net nothing. Oddly, the notion that all the fundamental properties of the universe add up to zero comes as no surprise, either. Physicists have long known that the net electric charge of nearly everything is close to zero. A comet or a cloud or even a cat may contain untold trillions of positively and negatively charged particles. But because the pluses and minus neatly cancel, the sum is very close to zero. Likewise, the universe as a whole appears to add up to nothing, electric charge–wise.

The same is true of every other fundamental property. All the spins of all the particles and all the stars and planets and galaxies are precisely canceled by equal and opposite counterspins. The net "spin" of the universe and everything in it is zero. And so it goes. Every particle is created along with an exact mirror-image opposite, or antiparticle. Despite all the universe does and all it's been through, it all adds up to a lot of nothing, always has, always will.

Nothing—unlike diamonds—really may be forever.

There was only problem with this picture: Where did all the matter and energy come from? There's no obvious source of negative matter or energy in the universe to add to the positive stuff to get zero. (Antimatter is the mirror image of "normal" matter in other respects, but it still has positive energy.) "There are something like ten million million million million million million million million million million million million million million particles in the region of the universe that we can observe," Stephen Hawking observes. "Where did they come from?"

The particles can be created out of the vacuum, given sufficient energy. But what was the source of the energy?

The answer that physicists like to give goes something like this: The energy comes from the "gravitational potential energy" of the universe. Here's one way to think about it. If you lift a heavy rock off the ground, the energy you put into lifting gets effectively stored in the height. If you drop the rock, it gains back that energy—from gravity—as it smashes to the ground. Even birds are smart enough to learn this trick. Seagulls routinely smash shellfish on rocks, using the potential energy of gravity to pry loose their supper.

Physicists count this stored energy as negative. Some will tell you the negative sign is by convention. Others will argue that it's negative because you have to put energy in (as in lifting the rock) to create the gravity field—and therefore it's negative. I urge the reader not to worry if this doesn't make complete sense.*

When confronted with difficult ideas like negative energy, I like to remember some sound advice I got very early on from Victor Weisskopf of MIT. I was frustrated over my attempts to really understand gravity as the curvature of spacetime, and Viki, after telling me he sometimes felt a similar frustration, passed along the following fable—a good story to keep in mind in all that follows:

A peasant asks an engineer to explain how a steam engine works. The engineer gives a detailed explanation, drawing diagrams, explaining fundamental concepts, showing where the fuel goes in and the steam goes out, how heat is transformed into motion, and so forth. When the engineer is finished, the peasant says: "Now I understand perfectly. But where is the horse?"

And that, Viki told me, is how he felt about Einstein's general relativity. "I understand it perfectly. But I don't know where the horse is."

In much of what follows, the reader, too, may be hard-pressed to find the horse. But like the steam engine, these theories pull enormous weight; they have turned out to be powerful, predictive, precise. In short, they work.

And once you accept that gravitational energy is negative, it's fairly easy to accept that a whole universe of matter and energy can be created from nothing—so long as an equal amount of negative gravitational energy is created at the same time. The account books are balanced. The universe is one big zero.

"The universe," concludes Guth, "could have evolved from absolutely nothing in a manner consistent with all known conservation laws."

---

*Alan Guth makes an excellent attempt in appendix A of *The Inflationary Universe: The Quest for a New Theory of Cosmic Origins.*

NOTHING BLOWS UP

The classic big bang theory describes the aftermath of the bang, but makes no attempt to describe what "banged," how it "banged," or what caused it to "bang."
—ALAN GUTH, *The Inflationary Universe*

How did a baby tweak in the vacuum grow up to be the whole universe? For a long time, no one could answer this question. While the big bang seemed to stand as the only possible explanation behind the expansion of the galaxies and the fossil radiation, it created almost as many problems as it solved.

For example: The universe is simply too big to be so smooth and uniform as it appears. In several important respects, the universe is uncannily—unbelievably—homogeneous. Everywhere you look, you find more or less the same amount of matter and light, behaving in more or less the same way, assembling into the same formations. It's hard to believe that things fell in line that way by accident. Everything in the universe behaves as if it were part of a 13-billion-light-year-wide marching band—stepping in precision and keeping its lines perfectly straight even though the marchers at the far ends can't see or talk with one another. In fact, the opposite ends of the universe are much too far apart to communicate in any way what they've been doing since the big bang—even at the speed of light. And according to the signs written into the fossil radiation, they always were.

So how did everything get so straight, so smooth, so synchronized?

A lucky accident, perhaps? What are the chances? Guth compares the lucky accident scenario to finding a stone the size of a marble so perfectly round that no bump on its surface steps out of line by a distance as large as a quarter wavelength of visible light. "If such a stone were somehow found," he writes, "I am confident that we would not accept an explanation of its origin which simply proposed that the stone started out perfectly round."

Something happened to make the universe so smooth. But what?

In the process of trying to solve a seemingly unrelated problem, Guth found the answer. The universe, it seems, inflated suddenly, unreasonably, in a manner truly beyond imagining. The newborn speck of nothing inflated in a tiny fraction of a second so fast that the numbers don't begin to make sense to our minds. It inflated much faster than the speed of light. (Nothing can travel faster than light—except space itself, which can.) To be precise, it grew so big that its size increased by a number followed by 50 zeros (remember that a million has a puny 6). It did that in a fraction of a second so small that to write it you'd have to put a 1 in the numerator and a number with 33 zeros in the denominator.

At the end of this exponential growth spurt, the baby universe was not really what you'd call big—only about the size of a cantaloupe. But inflation moved fast enough to ensure that all corners of the universe would have been in touch for sufficient time to synchronize the patterns that would subsequently form. Moreover, the explosive expansion smoothed out any innate crookedness. Take any wrinkled balloon and blow it up—big. Irregularities automatically iron out. Like a plastic surgeon taking the knife to the wrinkled face of an aging actor, the greater the stretch, the smoother the surface gets. Inflation effectively gave the universe a face-lift. (This may explain why cosmologists are occasionally confused with cosmetologists.)

What's more, any surface of any shape that blows up large enough looks flat on a small scale—just as our Earth is effectively flat over the span of your own backyard. One of the more unsettling implications of inflation is that our universe is just such a tiny patch on a larger stage (more on that later).

Inflation did more than iron out the wrinkles in the big bang. It also explained where all the structure in the universe came from. Because, for all its endless uniformity, the universe doesn't just sit there. The band members kick up their heels. Matter falls into gravity pits and ignites into stars, galaxies drag stars around in great gravitational whirlpools, gravity grabs clusters of galaxies and arranges them in long strings and sheets that stretch hundreds of millions of light-years across the sky. If everything had started out perfectly the same—a blank sheet of nothing—none of this could have come into being. Somewhere, a seed had to help things clump. Somewhere, a tiny excess of energy had to warp spacetime into a well deep enough to start to pull other stuff in.

What could have made those first clumps? The answer is as simple as it is surprising: The first irregularities were created by the uncertain jittering of the newborn speck of nothing itself. When the quantum fluctuation that started the universe blew up, it took all its uncertainty with it. And each one of those jitters exploded into a warp in spacetime that today guides the paths of giant clusters of galaxies. Those initial wiggles, in other words, expanded to form the scaffolding that even today supports everything in the universe. Without them, nothing would have stayed not much of anything.

THE DYNAMITE VACUUM

If meringue is made by beating egg whites and sugar, how do you make a universe? . . . In contrast to the big bang recipe, the inflationary version calls for only a single ingredient: a region of false vacuum.

—ALAN GUTH, *The Inflationary Universe*

What powered this incredible inflation? Neither horse nor steam engine, to be sure. Nothing existed except the tiniest speck of nothing. But that speck, it turns out, was packed with power. In fact, the nothing that tweaked to create our universe wasn't a real nothing at all. It was an impostor, a fraud, a charlatan. It was worse than nothing; not even nothing. It was, in short, what is known as a "false" vacuum—and for good reason. The false vacuum is a vacuum that lacks the most critical requirement for any self-respecting empty state—an absence of energy.

What qualifies this as a vacuum at all? A physicist would say that the false vacuum is the lowest possible available vacuum under present circumstances. It's as low as you can go. This false vacuum is something like a stick of dynamite lying on a table, to use Jan Rafelski's example. It doesn't do much, but it harbors great hidden reserves of energy. Light a match, and it explodes.

Think of a ball perched on a shelf, like the apple caught in the bird's nest mentioned in Chapter 4. Most of the time, the shelf is the ball's lowest possible energy state. A spider, say, making a home in the corner of the shelf could easily be deceived into thinking that the ball is as low as it can possibly go. The smallest jolt, however, could send the ball crashing to the floor—releasing a great deal of energy in the process.

The shelf is a better analogy than the dynamite, because it better reflects the many kinds of false vacuum states that physicists explore. For example, there are very stable false vacuums—analogous to a wide shelf with a deep groove running along it. It would take a very big jolt to jar the ball out of such a groove and onto the floor. But there are also highly unstable false vacuums—more analogous to a very narrow shelf where the ball perches oh so precariously; the merest breath of air will send it toppling. Then there are false vacuum states that change only very slowly, like very gentle inclines. Anything could get the ball rolling, but once it started, it could take years (or eons) to completely give up its energy to the floor.

Another common analogy—also encountered previously—views the false vacuum as a kind of water, vacuumy enough to any fish swimming around in it, but clearly not at its lowest energy state. When this water freezes (that is, falls to a lower energy state) it, too, releases energy, just like the falling ball.

Of course, the universe is not a stick of dynamite, a ball on a shelf, or an ice cube. The universe is the universe. All these analogies are approximations, helpful in different contexts to different degrees.

No matter how you think of it, however (and even cosmologists don't agree), the universe got the energy to inflate when a piece of this false vacuum began to "fall" into a true vacuum state. The energy of the impostor nothing is what powers inflation. In fact, it only takes about an ounce of "false" vacuum to create the whole universe, according to Guth, who figured out how to do it. This vastly simplified the business of cooking up a universe. The standard big bang recipe calls for a hot soup of particles and energy smashed into a very small space—condensed soup, you might call it. The inflationary universe calls for only a region of highly condensed "false" nothing.

Like most good theories, inflation solved some problems, and created others. For example, inflation happened so fast and its effects were so thorough that it wiped the slate of the universe clean, leaving no traces of its former self. The face-lift was so extreme, in other words, that there was no way to reassemble the original appearance of the universe before the operation. Whatever was happening in the universe before this explosive event—whatever particles, structure, or energy may have souped up the universe in its moment of conception—all this is gone in a poof of rapid expansion. Inflation is now the starting point for everything else.

Inflation, in effect, set up the universe for the big bang. But here, too, lies a problem. Just as compressing things into a small space tends to heat them up, so letting them expand tends to cool them down. Pump up a tire and it gets hot; conversely, rapid expansion of gases produces the "frig" in refrigeration. But if inflation stops cold, how did the universe begin with a fiery hot bang?

PARTICLES OF NOTHING

Somehow, the energy is extracted from the vacuum and turned into particles. . . . Don't try it in your basement, but you can do it.
—cosmologist ROCKY KOLB, University of Chicago

To talk about the "pure energy" of the vacuum doesn't mean anything. Energy has to have some form. That form is all the particles that could be in our universe.
—physicist HELEN QUINN, Stanford Linear Accelerator Center

So here we have the universe, ready to be born—a cold, puffed-up piece of nothing. How do you get from there to here? Where does all the *something* come from that forms the universe we see today?

During inflation, there are no particles; there are not even any fields, except the one that describes the false vacuum as it falls off the shelf from one state to another. Physicists don't understand this field very well, but they have a name for it nonetheless: the inflaton field. At the end of inflation, the inflaton field is the whole of everything. All the energy that went into propelling its explosive growth is now stored inside—like the energy stored in a boulder pushed to the top of a cliff. Maybe it will come straight down. Maybe it will land on an outcropping of rock and hang out awhile before releasing its full punch on some unwary passerby below. Either way, what goes up comes down, and when the inflaton releases its energy, that energy becomes everything.

The release of energy may explain how the big bang got hot in the first place. Like water freezing into ice and releasing its energy into its surroundings, the "freezing" of the vacuum liberates enormous amounts of energy.

It's no real mystery how this energy turns into particles. As Helen Quinn points out, energy has to take some form. That form, in the quantum world, is particle-like. Particles, after all, are nothing but energy packaged in quantum mechanical lumps. So once you have energy, particles come naturally. The newly frozen structure of the vacuum determines what kinds of clumps can materialize. As simply as water freezing into ice, the inflated vacuum froze into the structure that gave rise to quarks, electrons, and eventually us.

It also froze into a lot of stuff we don't know about, some physicists believe. In many cosmic scenarios, dark matter particles produced during this freezing process played a major role in forming the original clumps that curved spacetime sufficiently to form large-scale structures like galaxy clusters. These particles may still be hanging around, playing some role behind the scenes. Obviously, these particles don't interact much with normal matter; if they did, they'd have been detected long ago. Instead, they go by the descriptive acronym WIMPS, "weakly interacting massive particles." But even WIMPS could pack a wallop. Kolb suggests some of the WIMPS created in the very early universe are so massive they should properly be called Wimpzillas.

None of this tells us, of course, what pushed the boulder off the cliff in the first place—that is, what cracked the mask of the false vacuum, setting free the real vacuum to show its true colors and create the universe. What happened to make the false vacuum blink? Maybe nothing. It's likely, many physicists believe, that the false vacuum was simply unstable to begin with; like the ball on the edge of the shelf, it didn't need much urging. Any random wiggle could have done it. And random wiggles are what vacuums—even false vacuums—are all about.

If true, the false nothing fell from grace and turned into everything because it was primed to do so from the outset. Or, as Frank Wilczek summed it up years ago in *Scientific American,* "The answer to the ancient question 'Why is there something rather than nothing' would then be that 'nothing' is unstable."

WHY MATTER?

Were it not for this imbalance, Genesis would
read differently, ending abruptly after
Let there be light and there was light.
—ANDREW LANGE

The creation of particles from energy seems straightforward. Energy comes in clumps. Pairs of particles and antiparticles appear out of the vacuum all the time. There's no mystery to this equation. $E = mc^2$ is known to every child in school.

The mystery is What happened to the antimatter?

When matter is created out of energy, it always comes in pairs: one matter particle set off by one antimatter particle, keeping the total amount of matter in the universe precisely zero. But something broke the rules, upset the balance, destroyed the perfect symmetry of nothing, and brought everything into being. That's why we're here. If matter and antimatter still filled the universe in equal numbers, each would have annihilated the other—transforming the mass of both into radiation. There would be light, certainly, but no cosmologists around to worry about where it came from. Something happened to give matter the upper hand.

And whatever happened, happened very fast. When the universe was born, the amount of matter and antimatter created had to balance exactly. But by a millionth of a second after the birth of the universe, the balance was already upset. By one millionth of a second after the birth of the universe, says Guth, quarks outnumbered antiquarks by a ratio of 300,000,000 to 299,999,999. Not much of a difference, perhaps. But a residue large enough to create the ingredients that went into making everything in the universe.

As to *what* exactly happened to create the excess of matter, the question presently stews on the front burner of science. But nature has not been stingy with hints. In 1964, physicists discovered that matter and antimatter are not always mirror images; in certain rare cases, a particle will transform into an antiparticle more often than vice versa. Today, several laboratories around the world have created "factories" designed to produce vast numbers of similar strangely acting particles. With luck, nature will be kind again, offering enough hints to piece together a solution.

But inflation certainly exacerbated the problem of trying to explain the existence of matter. The simple old big bang, after all, could have simply started from a hot, condensed soup containing just the right number of matter particles. But no matter what that speck of soup looked like, inflation would have diluted it into effectively nothing. Inflation means, among other things, that matter can't be taken for granted; its presence has to be explained.

Inflation raises a host of other thorny issues: for example, assuming there are extra dimensions, why did some dimensions inflate, while others remained curled up and tiny?

And it doesn't, ultimately, provide an escape from the chaos that ensues when extreme gravity squeezes into close enough quarters to get tossed about by the quantum uncertainty. As we've seen, under such conditions, the natural jitter of the quantum gets so big that space and time dissolve into a kind of foam. The size of the universe approaches zero, but the curvature of spacetime and the density of the energy become infinite. And whatever led to inflation in the first place undoubtedly was governed by the laws that rule this realm. Since no one knows the laws, many physicists would rather not go there. In fact, some will argue that any discussion of what happened prior to inflation is beyond the realm of real science. At least for now.

NO BOUNDARIES TO NOTHING

I think that imaginary time will come to seem as natural as a round earth does now.
—STEPHEN HAWKING, *Black Holes and Baby Universes*

There are ways to avoid a descent into the hell of quantum gravitational chaos. Stephen Hawking and colleagues, for example, have proposed a quite different approach that makes a neat detour around the notion that the whole universe can be condensed into a single point. There is no single point, they say, because the universe is something like a spherical earth. The farthermost north point on the globe is the north pole. We don't need to worry about what is north of the north pole, or what "created" the north pole. It is simply there. To get from there to here, we simply start walking.

In Hawking's scenario, the north pole is analogous to the beginning of time. The universe is analogous to a four-dimensional sphere. Increasing latitude represents the passage of time. The circumference of the sphere at any latitude represents the size of the universe. As the universe grows from a point (north pole) to maximum expansion (equator), it evolves into the cosmos we see today. But there is nothing intrinsically special about the point we call "north pole," beyond the fact that it is our starting point.

There is one catch: The time that Hawking uses is "imaginary" time. It's a lot less mysterious than it sounds. Imaginary numbers, you may remember, are those that result from trying to take the square root of a negative number—for example, the square root of –4 is $2i$. Mathematicians called them "imaginary" because they don't exist in the normal number system; normally, you can't take the square root of a negative number, because the product of two

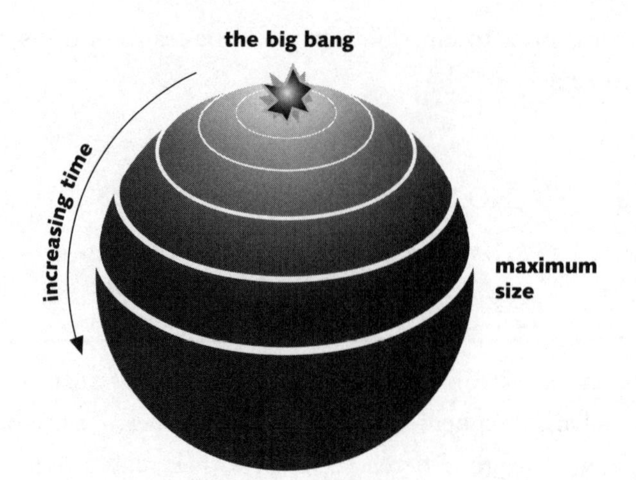

No time before the big bang; no "north" of the North Pole.

negatives is always a positive. (If –2 × –2 = 4, how could you ever have such a number as the square root of –4?) The only way to get a negative square is to start with an imaginary number; then, by definition, the square is negative.

In any event, using imaginary numbers allows physicists to plot time as if it were just another dimension of space. If you have a three-dimensional axis, with *x, y,* and *z* perpendicular coordinates, you simply add a fourth perpendicular coordinate for the imaginary numbers. That coordinate denotes the passage of time.

So what? you might well ask. What difference does it make whether time is imaginary or "real"? The critical difference is that real time moves inexorably backward to a point where the universe collapses into a singularity. But imaginary time doesn't come to a dead end. Instead, it can stretch infinitely, just like dimensions of space.

"Instead of time being real, it wanders around in the realm of imaginary numbers," said Caltech physicist John Preskill. "Now it looks a lot like space. Now it can go both ways."

Can physicists get away with this trickery? Is it fair? Most would argue that imaginary time—just like real time—is merely a convention that happens to work because it accurately describes reality. So it really doesn't much matter which time we use, any more than it matters whether you say that Earth is flat (from your backyard) or round (from space). It's simply a question of perspective. As Hawking argues, both versions are simply models that physicists made up. "So it is meaningless to ask: Which is real, 'real' or 'imaginary' time," he says. "It is simply a matter of which is the more useful description."

Hawking's approach not only avoids the dreaded singularity at the center of the big bang, it also deftly avoids the problem of where space and time came from. If you view the birth of the universe as a fluctuation in some preexisting vacuum, then space and time already had to exist. A fluctuation, by definition, is something that changes in space and time. So space and time have to exist before the fluctuation. And space and time, as we have seen, are not nothing. Even empty space is real, and its presence has to be accounted for just like everything else. "In this context, a proposal that the universe was created from empty space seems no more fundamental than a proposal that the universe was spawned by a piece of rubber," writes Guth. "It might be true, but one would still want to know where the piece of rubber came from."

In Hawking's scenario, the universe doesn't have to be created from nothing. It's just there. It is a geometry of space and time. When the geometry changes—that is, when imaginary time becomes real time—the universe begins. The change from nothing to something is literally a change in geometry.

"I wouldn't describe it as a universe being created from nothing," Hawking elaborated at a recent press briefing at Caltech. "Instead, the universe just exists as closed Euclidean geometry. It doesn't have time so it doesn't have a beginning or end. The time we experience is something we construct."

The big bang happens, he said, when you start moving in the imaginary direction from any point on this sphere. Or in the words of Hawking's collaborator, Neil Turok of Cambridge, "Where space turns into time, you could say time began."

THE RETURN OF FOREVER

I believe that soon any cosmological theory that does not lead to the eternal reproduction of universes will be considered as unimaginable as a species of bacteria that cannot reproduce.

—ALAN GUTH, *The Inflationary Universe*

These days, the big bang is a given. So is inflation. So is the idea that the universe amounts to nothing, and could have originated from nothing. But cosmologists continue to debate the mechanism that created inflation in the first place. Quantum fluctuation in an unstable vacuum is but one idea among many. Yet pervading nearly all the scenarios for the birth of the universe is a sense that our universe is not alone. We are not one of a kind; more likely, we are one of a litter. Universes, it turns out, may well reproduce like rabbits.

Guth, for example, doesn't see any reason why the false vacuum can't just keep on creating new universes. Every time a speck of false vacuum inflates into a real universe, the rest of the false vacuum surrounding it grows and grows. More and more nothing comes into being—and along with it more and more chances for a new universe like ours to sprout. We'll never run out of nothing. So we'll never run out of the raw material necessary to make a new universe.

Guth is the first to admit this scenario has a starkly emotional appeal—much the same appeal, in fact, as the original eternal universe that was taken for granted by Einstein and everyone else. The big bang effectively erased the idea of "forever." Time and the universe came into existence, and would someday cease. Continually reproducing universes, on the other hand, are here to stay.

"I claim that once the viability of eternal inflation fully penetrates our psyches," Guth writes, "our mindset concerning theories of the universe will be radically altered. Today, the dominant view of the origin of the universe, in both Judeo-Christian and scientific contexts, portrays it as a unique event. . . . However, if the ideas of eternal inflation are correct, then the big bang was not a singular act of creation, but was more like the biological process of cell division."

The universe, in other words, is organic. The birth of the universe is a natural event. The universe is a process, not a thing. It just keeps right on happening. It may never have started, and it will almost certainly never stop.

This eternally self-reproducing universe could even explain in a natural way where our universe came from: its parent universe. According to Guth and others, a universe that breeds like rabbits is far more natural than a universe that simply pops into existence out of nothing. After all, he says, if you came upon a new species of rabbit in the forest, and you had to figure out where it came from, you could guess that it spontaneously assembled out of random molecules or was created by some other mysterious cosmic event. More likely, you would conclude that the rabbit was produced in the normal way—by other rabbits.

"When we notice that there is a universe and ask how it originated," he writes, "the same inferences that we made for the rabbit question should apply."

Guth is certainly not the only scientist to explore the notion of multiple universes. Smolin, Hawking, and others see baby universes emerging out of the backsides of black holes. Stanford University's Andrei Linde envisions continually inflating universes as fractals—something like the continually branching limbs of a tree. The big bang, in his scenario, is not just one big fireball. "It consists of many inflating balls that produce new balls, which in turn produce more new balls, ad infinitum," says Linde. Each of these universes could have different laws of physics, different particles, different kinds of space and time. "The theory is very simple," he says, "but we have had a lot of psychological barriers to overcome."

And clearly, the psychological implications are enormous. If the greater universe (what do you call something larger than the universe?) is literally littered with universes, what makes us in any way special? Some chance combination of features, perhaps—gravity just the right strength, a vacuum of just the right structure, space of exactly the right dimensions, not too hot and not too cold, not too big and not too small. Like Goldilocks and baby bear, the fit between ourselves and our universe has to be just right.

From our parochial perspective, at least, our universe is the pick of the litter. At the same time, it is merely one of a huge cosmic ensemble.

IS IT SCIENCE?

Cosmology used to be considered a pseudoscience
and the preserve of physicists who might have
done some useful work in their earlier years,
but who had gone mystic in their dotage.
—STEPHEN HAWKING, *The Nature of Space and Time*

A universe that appears as an instanton in the shape of a wrinkled pea; eternally inflating nothing; self-reproducing universes—does any of this qualify as science?

The answer you get depends on whom you ask. Certainly, all of these ideas are logical consequences of well-tested theories. In that sense, alone, they qualify as science—if extremely speculative science.

But are the consequences themselves testable? Perhaps. The most important experiment is the one that already happened, some 13 billion years ago. The early universe left its smudgy fingerprints on the old light still streaming at us from the time that light and matter parted ways—the cosmic microwave background. As a new generation of satellites sets out to map this background in greater detail, they may send home at least some of the answers.

"We view the universe as a physics experiment," said Turok. "We're at the point where we're about to take the read-out of the universe."

On the theoretical side of things, the earliest moments of time present a stark challenge to string theorists. After all, a theory that describes everything should certainly be able to describe how everything came to be. A fully functioning string theory should be able to tell us not only the shape of the universe at its moment of birth, but also what principles determined that shape. If the universe sprouted from a speck of false vacuum, string theory should be able to say why that vacuum, and not others. Somehow, the false vacuum—like everything else—is ultimately created from strings. Strings, in other words, weave the conditions that made the universe ripe for the big bang, and everything that followed.

If the theory turns out to be right, it could offer another route beyond—or even around—the problems inherent in the big bang. Perhaps, suggests physicist Gabriele Veneziano of CERN, the big bang is but one of many interesting phases in the life of the universe. Rather than the beginning of time, it's a turning point, like reaching twenty-one. The big bang occurred when the universe was at its maximum density, temperature, and curvature, but not necessarily at infinite density, temperature, and curvature. It was only going through a phase—albeit a traumatic one. The time before time began becomes "negative time." As you dial the clock backward in negative time, things don't come to a stop; instead they get less and less interesting until nothing happens at all.

"The Big Bang loses its historical meaning of initial time to become a more modest, though still important, turning point in the history of the universe," Veneziano writes in the *CERN Courier.*

There is, in fact, evidence afoot that the universe might just be gearing up for such another phase change. And what is the driving force behind the coming upheaval?

Why, nothing of course.

Chapter 8

NOTHING IN THE NEWS

One of the most surprising recent advances in cosmology is that 75% of the Universe seems to be made of nothing.
—CHARLES LINEWEAVER, *What is the Universe Made Of?*

IN 1998, NOTHING made the news in a big way. Not only was nothing on the front page of every major newspaper, it was even selected as *Science* magazine's "Breakthrough of the Year."

Nothing was so honored because it was caught in the act of pushing the stars around—or more precisely, creating more and more nothing in the space between galaxies, thereby causing galaxies at the edge of the universe to speed away from one another at ever-increasing speeds. The galaxies were behaving as if some repulsive force—some antigravity—were pushing them apart. The force appeared to be revving up the galaxies like an airplane gunning its engines. The galaxies were under the gun, and the gun was nothing.

This accelerating push of nothing couldn't be seen directly, of course. But galaxies bobbing around in space are swept along by the all-pervading emptiness like twigs in a stream. So by measuring how fast these distant galaxies are speeding away from us, astronomers can indirectly gauge the speed of the expansion of space itself.

In truth, the measurement is even more indirect than that. To track really distant galaxies, astronomers need a bright beacon of some sort—preferably one that sends a consistent signal, no matter where it happens to be in the cosmos. As a stand-in for galaxies, astronomers use a type of supernova, or exploding star, that lies within the galaxy. These particular supernovas appear to brighten and dim in consistent patterns. Like fireflies blinking love calls to potential mates, the exploding stars light up in ways that tell astronomers something about their intrinsic properties—in particular, their innate brightness. Using that brightness as a standard yardstick, astronomers can compare it to the measured brightness of the supernova to figure out the star's true distance. The dimmer the light, the farther the star, the more distant the galaxy.

Measuring anything at the far edge of the universe needs to be approached with a certain amount of caution, to be sure. Things get in the way—dust, for example—and the supernovas themselves are not well understood. There is a real danger, as Rocky Kolb has put it, that scientists might be engaging in the sin of premature consensus. (This becomes especially tempting when the observations of accelerating galaxies seem to be just what the cosmologists ordered, as we shall see.)

But while the evidence is still highly controversial, most physicists feel that the accelerated expansion of empty space is a real—and troubling—possibility. And that acceleration, in turn, is the "smoking gun," as Michael Turner calls it, revealing that a strange new antigravitational force seems to be at work. "If the universe is really accelerating, then we have something out there with negative pressure," Turner said, "and that's something new."

NOTHING GETS A BAD NAME—AGAIN

I am a detective in search of a criminal—
the cosmological constant.
—SIR ARTHUR EDDINGTON

Nothing was new, in 1998, about simply more and more nothing. Astronomers had known since Hubble that the universe was expanding and had been expanding ever since its explosive birth in the big bang. The amount of emptiness in the universe had been increasing since the beginning of time.

However, until 1998, the expansion of emptiness appeared to be reined in by the counterpull of gravity. In fact, the outward momentum of expansion and the inward pull of the mutual attraction of everything in the universe seemed to balance with remarkable precision. Gravity and expansion were equal adversaries in an ongoing tug of war, locked in a seemingly permanent stalemate.

Or more poetically: Think of a galaxy flying through space as a ballerina taking off on a giant leap across the stage. The force of her muscles propels her upward, just as the explosive force of the big bang propelled everything in the universe outward. But at the same time, gravity keeps trying to pull the ballerina back to earth, just as gravity pulls all the matter in the universe back toward a common center. If the ballerina jumps too high, too fast, she could defy gravity and leave the earth for good. If her muscles are too weak, she'll fall on her face before ever getting very high off the ground. But if the force of her leap and the pull of gravity balance just right, she'll stay precariously balanced exactly between those two options.

Just so, the universe seemed to float on the delicate boundary between expansion and contraction, the push of the big bang and the pull of everything.

The 1998 discovery meant, in effect, that the ballerina was receiving a slight push from some unknown force, some mysterious wind beneath her wings taking her upward faster and faster, perhaps never to return to earth. The galaxies, like the ballerina, were taking off for parts unknown. Expansion had won the tug-of-war.

What kind of repulsive force could give expansion such an edge? Actually, Einstein had inserted just such a repulsive force into his early theory of gravity—by necessity. At the time, no one knew that the universe was expanding. Like everyone else, Einstein believed the universe to be static, unchanging. And yet, his equations said it could not be so. They showed that gravity acting alone would have long ago collapsed everything in the universe to a single point. In the absence of natural expansion set in motion by the big bang, Einstein needed an expansive force. He called it the *cosmological constant.* And when Hubble later discovered the expansion of the universe, Einstein called his own invention the biggest blunder of his life. Or as he wrote to a friend after hearing Hubble's news: "Then away with the cosmological term!"

The cosmological constant fell immediately into disrepute, rather like the ether. The repulsive force seemed best forgotten—a now-embarrassing fairy tale created to ease discomfort in a confusing world, something like the tooth fairy.

"It's the most maligned constant in the history of physics," said University of Chicago astrophysicist Joshua Frieman, one of the organizers of a meeting on the cosmological constant held at Fermilab after the accelerating galaxies were discovered. Frieman dug out some early reflections on the ill-fated repulsive force by Einstein's contemporary, Sir Arthur Eddington. Eddington went so far as to say: "If ever the theory of relativity falls into disrepute, the cosmological constant will be the last stronghold to collapse."

And yet, today cosmologists regard the return of the cosmological constant as a godsend. To them, the news of accelerating galaxies couldn't have come at a better time. In fact, even before the 1998 evidence, many were warming up to the idea that a cosmological constant might be just what was needed to repair certain gross inconsistencies in the shape of the cosmos.

NOTHING IN THE BALANCE

For the first time we have a full plausible accounting for all the matter and energy in the universe.
—MICHAEL TURNER

Cosmologists needed the energy of the repulsive force to balance the books of the universe in ways that agree with other observations. "[It] seemed to be just what was needed," said Turner. According to Turner and most of his colleagues, in fact, the ever-increasing energy of empty space may make up the vast majority of energy in the universe—some 70 percent.

The reason has to do with the fact that the universe appears to be almost uncannily "flat." As we have seen, space has shape, and that shape is determined by the amount of matter sitting within it. A universe with an abundance of matter would bend space so drastically that it would collapse in on itself, imploding before it ever had a chance to evolve. Conversely, a universe light on matter would have become so dilute so fast, so spread out, that no stars or galaxies or planets could have formed. But the universe seems to be just right: not curling in on itself; not curving outward; but almost perfectly flat.*

However, the universe does not appear to have enough matter to accomplish this feat. When physicists add up all the matter and energy in the cosmos, they find it lacking, and by a huge amount.

"We haven't weighed the universe with any precision," said Turner. But there is a good case to be made, he said, that the total

---

*See Chapter 5, "Nothing *Becomes* Center Stage."

amount of matter and energy is less than 35 percent of what it should be to keep the universe balanced.

There are (at least) two obvious questions here: Why do cosmologists think the universe is flat? And how do they add up all the matter in the universe?

The first question is easy: It's flat primarily because it looks flat. The first light coming at us from the earliest moments of "seeable" time doesn't bend severely enough to suggest a very curved universe. Moreover, if the universe really did inflate exponentially soon after (or before) its birth—and it gives every appearance of having done just that—then it would have to be flat today. Inflation flattens everything.

And a flat universe, in turn, requires enough energy and matter to keep up its end of the ongoing tug-of-war with expansion.

As to how astrophysicists know the amount of energy and matter in the cosmos, they use several different methods—all of which, rather remarkably, agree. For one thing, they look at the motions of stars and galaxies. Since the amount of matter and energy determines the curvature of spacetime, and the curvature determines the paths of galaxies and stars, those motions serve as a direct measure of the amount of matter and energy around. And the bottom line is, there isn't enough.

For another, the astrophysical understanding of the first moments of time is now solid enough for theorists to cook up all matter in the universe from scratch. Beginning with the big bang, they've predicted the exact amounts of hydrogen, helium, and lithium that should have been produced. When they look out into the universe, that's what they see. Using these now fairly well-understood recipes for matter creation, they can also infer the amount of dark matter necessary to hold clusters of galaxies together. Observations suggest the dark matter outweighs the visible matter by about 5 to 1. But still, it isn't enough to fight expansion.

About 70 percent of the necessary matter and energy is still "missing." No wonder cosmologists were so happy to see the cosmological constant appearing once again over their horizon. Perhaps the energy of this repulsive empty space would provide the necessary heft to keep the universe in balance.

The accelerating universe also promised to explain an unrelated problem: why the universe seemed too young for some of its own stars. Astronomers had determined that the oldest stars in the universe have been evolving for at least 15 billion years. (Astronomers judge the age of stars by looking at the elements they've burned—something like horse traders looking at an animal's teeth.) But if the universe had been expanding at roughly the same pace ever since the rapid burst of inflation stopped, extrapolating back in time only buys the universe about 13 billion years. How can the universe be younger than the stars it contains?*

The paradox would disappear if the universe was expanding faster now than it was yesterday. Imagine a car cruising from Los Angeles to New York at 50 miles per hour, said Case Western University physicist Lawrence Krauss, one of the early advocates for a return of the cosmological constant. Let's say the distance it travels is 3,000 miles. In that case, it would take 60 hours to reach New York. If you knew how fast the car was traveling, and you knew what time it arrived in New York, you could easily figure out what time it left L.A.

Unless its speed wasn't constant after all. What if the car traveled 50 miles per hour for the last part of the trip, but dawdled earlier in its journey? Then it would have taken much longer to reach its destination. In the same way, a universe that was accelerating could have taken much longer to reach the size it is today—giving

*These numbers are undergoing almost continual fine-tuning as experiments bring in new data.

even the oldest stars plenty of time to evolve to their present aged state.

NOTHING GETS REPULSIVE

This is desperation. The pieces aren't fitting,
and we're trying to make them fit. It's like you
understand how the heart works and how the liver works
and you try to figure out how the whole thing works.
For now, we have too many legs and not enough arms.
So, like Frankenstein, we sent Igor out into the graveyard
of old ideas to see what he can dig up.
—ROCKY KOLB

Calling the repulsive force the cosmological constant is one thing. Figuring out exactly what it is presents another problem entirely. As of this writing, there's so much confusion about the nature of the repulsive force that it's spawned a whole new vocabulary. This is not surprising. Scientists pushing into unknown territory often find themselves at a loss for words. The more mysterious the emerging landscape, the further they must reach for appropriate language to describe it. Among the top ten terms for the stuff presented by Joshua Frieman at one meeting were X-matter, quintessence, NACHOS (not astrophysical compact halo objects), and Roll-ons. Michael Turner calls it simply "smooth stuff," or sometimes "funny energy." Others return to Einstein's nomenclature and call it the cosmological constant.

One thing it's definitely *not* is antigravity. Gravity is the mutual attraction of matter and energy. Antigravity would imply that matter and energy have somehow become repellent, repulsive. However, the repulsive force in the universe doesn't come from matter and energy. It comes from empty space itself. Rather than pushing on matter, it stretches the space in between bits of matter, adding more and more nothing. And that more and more nothing adds up to an outward push.

Physicists describe this negative force as pressure, like the pressure that pushes out from the inside of a full balloon, except that this pressure actually *pulls.* It pulls in a direction opposite from gravity. Why is it negative? Why does it defy gravity? "The only honest answer to this question is that the relevant equations say so," concludes German physicist Henning Genz in his book *Nothingness: The Science of Empty Space.* "A lot of scientific and popular writers' ink to the contrary, there is no better explanation. The equations that say so are as compelling as those that make two plus two equal to four."

Other physicists say much the same thing. "I don't know a good heuristic argument about why it should be so crazy," said Lawrence Krauss. "Why the energy that comes out of nothing is so different from the energy that comes out of something." Like the negative nature of gravitational potential energy described in the last chapter, you can logically talk yourself into believing in the negative pressure of the vacuum, but you might end up wondering: Where is the horse?*

The important thing is that the repulsive force comes from empty space itself; that means it grows with distance. Think of a universe full of matter, expanding and creating more and more nothing in the spaces between. As the universe evolves, the amount of matter (and energy) stays the same. But the amount of empty space increases without limit. Like the broomsticks in the "Sorcerer's Apprentice," nothing just keeps on multiplying. So even though the repulsive force of emptiness is puny compared to gravity, given enough space, it always wins.

The fact that nothing is a long-distance runner explains why we don't see its effects on small scales. It doesn't push Earth away from

*I have yet to find an understandable explanation for this phenomenon. Perhaps Lineweaver sums it up best when he writes: "It's a bit like discovering compressed springs everywhere in the vacuum of space. These springs make the universe expand."

the Sun or even the Sun out of the galaxy. There isn't enough empty space in those comparatively tiny regions to make a difference. But give it enough space for a playing field, and it can push galaxies right off the edge of the universe.

This also poses a puzzle: If matter gets more and more dilute as the universe gets bigger—while the energy of nothing gets stronger with the ever-increasing volume of empty space—then there will be only a very brief moment when the two forces are in exact balance. That moment appears to be now.

That is, it seems we currently live in a very special era, when the "funny" repulsive energy and the normal gravitational energy just about balance. Soon they will be out of balance again. Is it possible that life can evolve only in such a brief window of barely perceptible acceleration? Or is there another—less parochial—explanation for this odd confluence of events? (See below.)

Finally, the repulsive force is unlike gravity in that it's the same everywhere. Gravity increases where matter is concentrated (by definition), but the repulsive force is spread smoothly throughout space. This evenhandedness is the "constant" in Einstein's famous fudge factor. The energy of the cosmological constant never varies, and it can't clump. In fact, it can't interact directly with matter at all. Like the ether before it, the funny energy can't slow down the stars; it has to allow everything to flow through without the slightest interference.

THE UNREASONABLE LIGHTNESS OF NOTHING

This must be the worst failure of an order of magnitude estimate in the history of science.

—STEVEN WEINBERG, *Dreams of a Final Theory*

The most obvious wellspring for the strange repulsive force is the vacuum itself. In fact, standard theories describing the physics of the vacuum produce just such a cosmological constant. There is, however, a problem: the repulsive force created by known theories is much too big. Indeed, if the vacuum contained all the energy particles physicists expect it to, it would be so repulsive that you wouldn't be able to see your hand in front of your nose. Even at the speed of light, the light from your hand wouldn't have time to reach your eyes before the expanding universe pulled it away. "The fact that you can see anything at all," says Krauss, "means that the energy of empty space cannot be large."

It's not that hard to add up all the energy contained in empty space—repulsive or not. Empty space, as we have seen before, is a hyperactive player, a prolific producer of jittering fields and virtual particles. The sum of all the possible jitters and particles is infinite. And even when physicists tame the infinities with various mathematical tricks, the answer they get is still huge—about 120 orders of magnitude too big.

"Everything jumps in and out of the vacuum, including things we haven't discovered yet," said Andrew Strominger. "And they all contribute to the cosmological constant."

That, at least, is the answer theory gives. The only real way to know the weight of nothing is to measure it experimentally, and the only true measure of weight is gravity. If something has matter or energy, gravity will respond by warping; gravity *is* the warping; if there's no warp, there's no matter or energy.

So what does gravity "see" when it looks out into this vast writhing vacuum? Nothing. Or almost nothing. For some unknown reason, all the activity of the vacuum is completely invisible to gravity. And yet, each jitter of the vacuum, each pair of particles that pops in and out, should be creating a small gravity well around it, just like a rock or a planet. The combined warping of all the gravity wells should be huge. "You would think," said Strominger, "that the whole universe would collapse because of the gravitational attraction of this sea of virtual particles."

There are ways to get the energy of the vacuum to disappear, but they are not worked out and tend to be fraught with problems. For example, as it turns out, different kinds of elementary particles add different kinds of energy to the vacuum: some contribute positively, some negatively. According to some theories, they could conceivably cancel each other exactly (another net zero).

If the galaxies really are accelerating, however, that only makes the problem worse. It would mean that the positive and negative contributions to the vacuum cancel out to 120 decimal places—but with a little negative pressure left over to push the stars around. It's much easier to get to exactly zero than to a tiny bit more or less than zero.

In short, the accelerating influence of nothing may have been great news for cosmologists. But it was very bad news for particle physicists, who are left trying to explain why all that energy adds up to zero or—even more puzzling—to just the tiniest smidgen more than zero.

NOTHING CHANGES

We want it to be here today and gone yesterday.

—Michael Turner

Aside from making the energy of nothing just the right size, physicists face a much bigger challenge: How to make that energy just the right size at the right place at the right time.

During the inflationary era, as we have seen, the power of nothing was huge: enough to inflate the cosmos exponentially from the size of a proton to the size of a cantaloupe in a few microseconds of time. Like the current acceleration, that expansion was propelled by the vacuum of empty space. How could the same force pack such a huge punch 13 billion years ago and be all but gone today? During inflation, the universe doubled in size something like a million times in a small fraction of a second; during the past 13 billion years, it's only doubled something like a hundred times. What damped it down?

Or as Weinberg puts it: "We want to explain why the effective cosmological constant is small now, *not* why it was always small."

Complicating matters further, the repulsive force had to be less when the galaxies were forming than it is today. If it had always hovered around its present strength, galaxies would have blown apart before they could have formed, and no one would be here today to worry about it. "We want it to be here today and gone yesterday, so that it doesn't interfere with the growth of structure," said Turner. This means the force had to start out huge, dissipate to almost nothing, and stay at nothing for about 13 billion years, then begin to rev up again ever so slightly in the not-too-distant past.

An inconstant constant? It seems to be the only possible solution. The force of the vacuum is with us at some times, but not at others, visiting and revisiting like a recurrent dream. But a constant that changes is by definition paradoxical and therefore messy.

It also implies, once again, that we are living at a very special time—basking in a brief interlude between enormous accelerations of nothing. And that's the kind of funny coincidence that sends physicists searching for explanations.

In any event, it's hard to imagine how the virtual goings-on in the vacuum could produce such a variable force. Instead, some physicists have proposed a new kind of funny stuff in the universe, called *quintessence.* The term comes from the fifth essence that ancient philosophers believed permeated the universe—in addition to the four fundamental essences of earth, air, fire, and water.

Quintessence would be an extra added ingredient in the vacuum, rather like the Higgs field. In fact, quintessence would have much in common with the Higgs field. It would be a scalar field, meaning it has strength, but no particular direction. It would be the same everywhere in space.

But it wouldn't be the same everywhere in time. It would be what physicists like to call a slow rolling scalar field—a sleepy beast that hibernates for long periods but has the potential to unleash enormous power. You can almost imagine this immense force of nothing exploding awake as the universe was born, only to fall back into a deep slumber for 13 billion years, and just now beginning once again to rumble under the covers, albeit in a much tamer form. Some physicists believe this field has always been around but was so swamped by the expansive force of the big bang that it's been effectively hidden. Now that gravity has slowed expansion down, the force of repulsion suddenly emerges from the noise. Whatever its real behavior, the negative power of quintessence would be fueled by the same negative gravitational energy that drove inflation.

Nothing begets nothing.

There are other candidates for the cosmological constant: knots in spacetime; wormholes that communicate with other universes; cosmic strings packed with vacuum energy left over from the early crystallization of the universe from hot bang to its present cool, habitable state.

"Quintessence," wrote Montaigne, "is no other than a quality of which we cannot by our own reason find out the cause."

BUT IS IT SCIENCE? REPRISE.

Let me remind you that there is
an answer to this, and nature has it.

—astrophysicist BILL PRESS, Harvard-Smithsonian Center for Astrophysics

We're entering a state of much more sophisticated confusion.

—CHRISTOPHER STUBBS

Of course, it is the role of science to find out why. But where to look for hard evidence of the nature of nothing? The beginning of everything, of course: the big bang.

A cosmological constant would alter ever so slightly the shape of the cosmic microwave background—the afterglow of the big bang. It would affect the distribution of the fluctuations in the radiation, the growth of those early ripples into large-scale structures, and the angles from which the wiggles appear to be coming. Like a lens with more glass bulging in the middle, a universe bulging with funny energy will bend light more (and in a different way) than a universe without.

It is a remarkable thing that this echo of such early times continues to reverberate through the cosmos, still carrying news of what happened before the universe was much of anything at all. In fact, the cosmic microwave background comes not directly from the big bang, but from some 300,000 years after. During those first 300,000 years, particles like protons were formed from quarks, but they were trapped in a hot soup so thick with radiation that you could not see through it even if you had been there to look. Electrically charged particles and light sloshed together like two mixed fluids, unable to disentangle from each other. The universe must have looked like the inside of a furnace—an opaque wall of light hiding everything that went before.

When things cooled down enough, negatively and positively charged particles could finally stick together without getting torn apart. Atoms were born. "Before that, the universe is very simple," as Caltech's Lange put it recently. "After that, we have atoms, chemistry, economics. Things go downhill pretty quickly."

The minute atoms were formed, the light was free to travel, and that's the light we see today—chilled to hundreds of degrees below zero. But each of those bits of light still carries the knowledge of what it was doing at the moment it gained its freedom. So the pattern of the sloshing is imprinted in the light. And just as water in a jar sloshes differently than water in an ocean—and mud sloshes differently than water—the nature of the sloshing will tell a great deal about the nature of the universe when it was 300,000 years young. If astronomers can read these barely discernible signals, they will have on their hands a clear snapshot of the shape, composition, and behavior of the very early universe. If the "funny energy" is there, it should stand out clearly.

There may be other ways to see it: When light from distant galaxies travels through large clumps of matter, gravity focuses it into multiple images. The size of the cosmological constant would affect how often such gravitational lensing occurs. The bigger the influence of the funny energy, the more empty space should fill the universe, the more galaxies should fill the space, and the more multiple images should be seen. Thus far, searches haven't turned up anything definitive, but the effort has really only just begun.

First and foremost, astronomers have to pin down the truth about the supernovas. Are the galaxies really accelerating? Are they really dimmer? Or are they just being viewed through a dark veil of intergalactic dust? The astronomers who made the measurements are certain they ruled out the most common source of pollution in the universe—red dust. But perhaps, as Press suggested, they've just discovered another kind of dust—maybe gray dust.

And then there's always the nagging question: Were supernovas different billions of years ago? Is there something about the stars themselves we don't understand? The particular species of supernova studied is thought to be fairly consistent because it always explodes near the critical point for supernova formation, which means it would always have the same mass and other characteristics. Other stars collapse and explode into supernovas when they get so massive that they can no longer resist gravity's pull. Any large-enough star can "go supernova" at the point when its internal nuclear fires run down and allow gravity to take over.

But this particular type of supernova begins the process as an already dead star. It's just sitting there in space, out of fuel, too small to collapse and explode. Somehow, astronomers believe, it sucks in extra matter from some neighboring star, or perhaps merges with a partner, and bingo! As soon as it's gained enough weight, it explodes. Given this scenario, all supernovas of this type should be roughly the same. The trouble is, no one understands the mechanics well enough yet to say whether or not the scenario is accurate.

In short, nature will clarify the mystery of the repulsive force and the acceleration of the galaxies. Astronomers have only to ask the right questions and learn to read the answers. In the meantime, many experimentalists and theorists alike will worry about jumping to conclusions. "It's such a big question, no one's going to completely believe it until they find another way to measure it," said Kolb. "You have to go beyond a reasonable doubt. You have to have more than a bloody glove. . . . The universe has tricked us before."

A STRINGY CONSTANT?

{The cosmological constant} is a monumental issue
not only for cosmology but for particle physics,
because it has to emerge from fundamental physics.
It has profound implications.

—PAUL STEINHARDT, Princeton University

As for theory, the cosmological constant poses perhaps the most interesting challenge of all. After all, the enormous energy of the vacuum is a product of that theory; if the theory is right, it has to be able to make the energy of the vacuum go away, gravitationally speaking. "Something is wrong," as Witten put it recently. "It's not just a detail. It's hinting that there's a mechanism in nature we don't understand."

For now, at least, many physicists are looking toward a savior that goes under the poetic name of supersymmetry, or SUSY for short. As the name explicitly implies, supersymmetry would suggest that the universe is far more symmetrical than it superficially appears. The universe we know is divided into two families of particles: the fermions, or (roughly speaking) matter particles, and the bosons, or (roughly speaking) force particles. They seem vastly different, but according to supersymmetry, each is part of a pair. Each fermion has a boson hidden in its history, and vice versa. These superpartners, or sparticles, as they are called, have not yet been seen, but they would have been present soon after the big bang. And physicists expect at least the lightest ones to make their appearance soon in accelerator laboratories.

Supersymmetry holds the promise of banishing the unreasonable weight of nothing because bosons and fermions cancel out, vacuumwise. What one contributes, the other takes away. So if they come in precisely equal and opposite numbers, the vacuum energy shrinks to nothing—just as gravity says it should.

Alas, while a perfectly symmetrical universe has no cosmological constant, we don't live in such a universe anymore. We never did. If the universe were perfectly symmetrical, we wouldn't be here. Somehow, the perfect balance that perhaps once existed between bosons and fermions was destroyed. And every explanation physicists have come up with for destroying the symmetry seems wrong because every one leaves us with a universe with a cosmological constant that is far too big.

"All known . . . mechanisms are wrong because they all produce a cosmological constant," said Witten.

Nevertheless, the problem looms not only as a challenge but as an irresistible lure. "The cosmological constant is a real obstacle to explaining a lot of things," says Witten, somewhat wistfully.

It is an especially tempting problem for string theorists. If string theory could come up with an answer to the cosmological constant, physicists would no longer have any excuse to disbelieve its power—lack of experimental proof notwithstanding. So, needless to say, the puzzle of the cosmological constant is the single question "stringy" physicists, as they are commonly called, would most like to solve.

"It's an embarrassment to all theoretical physics, and it's a particular embarrassment to string theorists because string theory is supposed to be a complete theory which describes everything," says Strominger rather typically. "It's the problem I'd most like to solve. I've tried everything I can think of and I don't have anything to show for it."

There are other theoretical approaches. Sidney Coleman's famous 1988 paper, "Why There Is Nothing Rather Than Something," proposes that quantum-sized wormholes connected to other (not necessarily real) universes could make the cosmological constant appear to disappear locally while maintaining its strength elsewhere.

(Forget Sagan. Maybe we already have contact.)

But the biggest challenge for theorists of all may simply be emptying the vacuum of all the trappings it's acquired over the past fifty years. "They've filled the vacuum with so much garbage, there isn't room for the cosmological constant," said Leon Lederman. "Einstein freed us from the ether. Now we need to get rid of [today's version of ether] again. We need to sweep the vacuum clean."

NOTHING IN THE FUTURE

We could be starting an inflationary era right now.
When Alan Greenspan reads about this,
the shit's going to hit the fan.
—ROCKY KOLB

Is history repeating itself? Are we gearing up for another period of inflation that will sweep away the universe as we know it? Possibly. The accelerating galaxies could be the tip of the iceberg—small twigs in a vast river of nothing that is just beginning to gather steam for a trip over the proverbial falls. If so, cosmic history could be repeating itself, 13 billion years after the fact. What comes around, goes around, including the power of nothing. "Inflation happened, and went away," said Steinhardt. "One question is Will this period of inflation end?"

No one will know until physicists better understand what forces wake the beast and what it takes to put it back to sleep again.

The good news is, such a new inflationary epoch would probably take a few billion years to get started. Then—when no one is around to notice—our galaxy might well be left an island in a vast empty space. As the expanding universe pulls all our cosmic neighbors out of sight, there will be no other galaxies to fill our telescopes, no neighboring worlds to scan for signs of probable life.

"If this is correct," said Kolb, "then in another 15 billion years, our galaxy will be a lonely place. A hundred years ago, we thought the universe consisted of our own galaxy. So in a 100 billion years, it will be only one galaxy again."

Chapter 9

NOTHING ON YOUR MIND

You ain't seen nothin' yet.
—AL JOLSON, quoted by Martin Gardner in
"More Ado about Nothing," *Mathematical Magic Show*

NONE OF US, if truth be told, have seen nothin' yet. That's because nothing is almost impossible to see. Like nature, the mind abhors a vacuum. Even in sleep, the brain ruminates, regurgitates, runs wild. It's a familiar story: Vacua can't stand still; black holes can't stop radiating, and human minds can't stop doing something, anything at all—which means that nothing is beyond the ken of everyday perception. "How can one describe nothingness, not-being, nonentity, when there is, literally, nothing to describe?" wonders neurologist Oliver Sacks in Richard Gregory's *Oxford Companion to the Mind.* Sacks's contribution to this compendium is an entry entitled "Nothingness."

Like a cloud, nothing has no edges. Ultimately, that's what makes the void so hard to grasp. It's not that it's not something. Rather, it has no firm boundaries, no place to hold on, no starting lines or entry points. The mind, like the hand, needs a handle. Where there is nothing to hold, we must create something to give the slippery emptiness some semblance of form. At the very least, we put a box around it or insert rulers and clocks to measure it. We must do something to nothing to get a grip.

This necessity for structure explains, among other things, why the mind creates concepts such as space and time. We can't think about anything—including nothing—without resorting to this minimal set of somethings. Space and time, according to eighteenth-century philosopher Immanuel Kant, come to us as "pure intuition." Kant was trying to demonstrate, Max Jammer points out in his classic, *Concepts of Space,* that "space and time are conditions under which sense perception operates."

Space and time, in other words, are the *sine qua non* of our mental landscape. You can't leave home without them. But they are only the bare requirements—the first spare steps. Upon that minimalist scaffolding, the mind erects and elaborates as necessary, filling up the void with sounds, sights, sensations, memories.

(And the mind not only creates something from nothing; when circumstances dictate, it can just as easily create nothing from something. See below.)

The mind's struggle to get a grip on perceptual nothings mirrors the efforts of physicists and mathematicians to deal with zero and the void. Each informs the other. To some extent, each is part of the other: The mind exists in a physical/mathematical universe. But that universe, in turn, ultimately makes itself known only within the confines of the mind.

Thus it seems only fair at this point to let the brain contemplate its own role in discovering and creating the emptiness we see in the outside world. In a sense, the holes in our heads are the windows to all other voids.

THE SOUND OF SILENCE

The idea of a void—of emptiness, nothingness,
spacelessness, placelessness, all such "lessness"—is
at once abhorrent and inconceivable; and yet it haunts
us in the strangest, most paradoxical way:
"Nothing is more real than nothing."
—OLIVER SACKS, in *The Oxford Companion to the Mind*

The perception of nothing is a vague concept at best. Probing its properties requires that we consider something concrete. Begin with the absence of sound—the sound of silence. Without the occasional respite of a pause, sound cannot carry meaning; syllables run together into noise; notes drone on.

The pauses themselves are players. Words unspoken are frequently as powerful—even more powerful—than words actually said. Playwrights such as Samuel Beckett and Harold Pinter are well-known for their sparse use of dialogue, for meaning carried mainly on long stretches of silence. The "Pinter pause" is as much a part of the playwright's vocabulary as his words, stage directions, characters, plot.

Silences speak volumes. This comes as no surprise. Many have felt impaled by the terror of the phone that doesn't ring.

"What's wrong?" she asks. "Nothing," he answers.

A rebuke? A reprimand? A pout? She'd better find out. Cordelia's "nothing" to Lear invited rage, jealousy, murder.

But nothing is not always tragic. Comedians couldn't make us laugh without it. That's what the ineffable quality called timing is all about.

In his essay, "Untitled: Art, Music, and Nothing," artist David Barker recalls the famous Jack Benny joke in which a robber tries to stick up Benny for his wallet. The robber demands: "Your money or your life!" Benny responds with a long silence. Finally, the impatient robber insists on an answer: "Well! Which is it? Your money or your life!"

Benny responds: "I'm thinking, I'm thinking."

It is Benny's silence, Barker tells us, and the exact duration of that silence, that makes the joke.

Silences get filled in as surely as black holes. Teachers who can bear three seconds of seemingly interminable silence in the classroom before rushing in to answer their own questions frequently discover that previously silent students begin to speak. Three seconds facing a class can seem like forever. Most teachers jump in to fill the gap before even a microsecond has passed, although most are unaware of their behavior. As it turns out, learning to live with dead air can provide just the space more thoughtful, generally quiet, students need to find a voice.

In music, of course, the rests are as important as the notes. "Notes are hung linearly on the clothesline of time," writes Barker. Before the twelfth century, composers had no notation for "rest"—no way to indicate an absence of sound; like mathematicians without a zero, their vocabulary was stilted, incomplete. Since then, however, precisely measured silence has become an essential compositional tool—a space to anticipate a shift in rhythm, build suspense, punctuate a thought like the period at the end of a sentence.

Taking silence to extremes, John Cage composed "4′33″" as four minutes and thirty-three seconds of silence—in three movements. The pianist sits facing the keys and does nothing. In the interim, sounds rush in: the breath and heartbeat of the audience, blood pulsing through veins; the unsmotherable sounds of the concert hall, the sighs and settling of the walls and floor.

(Cage's friend, the painter Robert Rauschenberg, created a series of plain white-painted canvases in the same spirit. Cage called them "airports for the lights, shadows, and particles." In another piece, Rauschenberg took a pencil drawing of Willem de Kooning and erased it, leaving only faint smudges as traces of what had been before.)

Not all silences are equal—not by a long shot. The tenor of the silence tells a great deal about the circumstances that make it possible. Like different physical vacua, different empty spaces allow different kinds of silences to sing.

At UCLA, Professor Seth Putterman presides over two totally silent rooms—huge vaults that stand as cathedrals to the study of sound. Pistol in hand, the physicist pries open a thick padded door leading to the sealed concrete chamber. Inside, the bare metallic-painted walls loom several stories high. He shuts the door, aims into the air, and fires.

Even though his firearm is only a high-caliber cap pistol, the shot sets off a thousand thunderclaps, the sound literally bouncing off the walls in endless echoes that reverberate deep in your bones. When it's finally over, and quiet returns, you can still hear the fluorescent lights buzz in your ears. Conversation is impossible as every word multiplies endlessly, a kaleidoscope of sound: the overlapping echoes turn words into jumbles of noise.

Putterman crosses a hallway, opens another enormous padded door into another empty space. This one is padded with deep wedges of foam. The floor is chicken wire: below, there is only air and more foam; the same above. As soon as Putterman shuts the door, a thick fog of silence descends, smothering even breath. Some people feel so claustrophobic they are forced to leave.

Once again, Putterman fires his gun at the air. A small flash of light sparks from the tip of the pistol. There is the slightest pop, like a wine cork gently leaving the bottle. The walls have swallowed the sound, stopped it dead.

"This is one of the leading dead rooms around." Putterman beams. (The first room is known around the lab as the "live" room.) What these rooms are alive and dead to, of course, is sound—which the rooms were specifically designed to amplify or squelch. Just as white light from the Sun looks red reflecting off an apple and green reflecting off leaves, so sound changes depending on what (if anything) is around to reflect the vibrations toward receptive ears. The smooth metallic walls of the live room bounce everything back. But nothing bounces off the walls of the dead room; all sound disappears into the foam's deep pockets.

The lesson is familiar: Depending on the character of the "nothing" you tweak, a very different kind of "something" emerges.

SEEING NOTHING I:
MAKING NOTHING OUT OF SOMETHING

We not only believe what we see:
to some extent we see what we believe.
—RICHARD GREGORY, *The Intelligent Eye*

The primary human sense, many would argue, is vision. The retina, in fact, is often considered a part of the brain. It is certainly the most studied instrument of perception, and as such has a lot to tell us about the holes in our heads, and the ways we see around them, or fill them in. While an in-depth look at visual perception is clearly outside the scope of this book, it is worth taking a quick tour of some of the ways the eye and brain conspire to make something of nothing—and vice versa.

Most familiar, perhaps, is our uncanny ability to erase signs that are right in front of our noses—including, for that matter, our noses, which always loom to some extent in our field of view. We don't see the blood vessels in our eyes, even though they are always to some extent obscuring our visual field. We don't normally see the distortions created by the curvature of the retina, the threadlike floaters that swim in the clear fluid of the eye's fishbowl, and a multitude of other physiological intruders.

The brain erases these sights because they are beside the point—adding no necessary data to the information we need to make sense of the visual world—and also because they are always present. Data that is always present is often perceived as "nothing": familiar sights and sounds, space and time, air and gravity, breathing and good health.

Filtering out extraneous information is something that the brain does superbly well. It has to. Without filtering mechanisms, the senses would flood the brain with far more information than it could sensibly use; most information that is out there, consequently, gets dumped long before it intrudes on consciousness. "We used to think perception was about going out and getting stimuli from the outside world," says neuroscientist Richard Brown, formerly of UC San Diego, now a scientist-in-residence at the Exploratorium in San Francisco. "Now we know it's mainly about filtering."

The eye comes naturally equipped with its own version of the elaborate "triggering" systems built into particle detectors. These computerized filters screen hundreds of thousands of stray ordinary unimportant particle tracks in order to focus on those few that may signal something new, something important. Without them, the detectors would be overwhelmed with irrelevant "noise."

(Curiously, even our memories must be continually filtered—and occasionally flushed—to prevent the clogging of perceptual pathways. Russian psychologist Alexander Luria wrote about a man who remembered too much, according to a recent exhibition on the subject of memory at the Exploratorium. The man could hold on to almost unlimited facts, as well as mental pictures of the people and places he had seen. "He was not a happy man," according to the exhibit. "His extraordinary memory for the minute details of the past overwhelmed and interfered with his experience of the present. Forgetting is clearly an important part of effective remembering.")

As necessary as this filtering may be, however, it certainly shuts our eyes to enormous amounts of otherwise readily available information. Perceptual psychologists call this often-unintentional erasure inattentional blindness. That is, we are blind to anything we don't pay attention to, which includes all that is familiar, and more. Drivers cruising down familiar streets often don't see road signs, children running into the street, spectacular sunsets. The information—the photons carrying the requisite frequencies and intensity of light—is present. Only attention is absent, asleep.

In one series of experiments by experimental psychologist Ronald Rensink and colleagues, viewers are shown rapidly alternating pictures in which something—usually something major—is changed from one frame to the next. Most viewers completely miss even the most obvious alterations. I, for one, failed to notice that a jet engine was missing from the picture, even though it took up a large chunk of the screen. In another set of images, the cabin of a boat comes and goes without making the slightest conscious impression. "Although people must look in order to see," the authors conclude, "looking by itself is not enough. . . . We argue that the key factor is attention." Inattentional blindness can turn something into nothing in very measurable ways.

(Needless to say, inattentional deafness is just as common a problem. How often have you talked with people only to discover they haven't heard a word you said? Or vice versa? Familiarity has the power to silence the singing of birds, the bonging of clocks, even the tooting of train whistles.)

Almost by definition, focusing on one thing fuzzes everything else into an invisible blur. Focus on a single face, and other people in a room can disappear. Focus on one voice at a party, and everyone else's words fade into background noise. Focus on a problem, and the rest of the universe can seem irrelevant, unreal. Perceptual illusions make the mechanisms behind these disappearing acts uncannily concrete. Consider, for example, the familiar illusion of the two facing profiles that become a vase, or vice versa, depending on which aspect of the image you pay attention to. If the vase is something, then the faces become background, or nothing. If the profiles are something, then the vase is nothing.

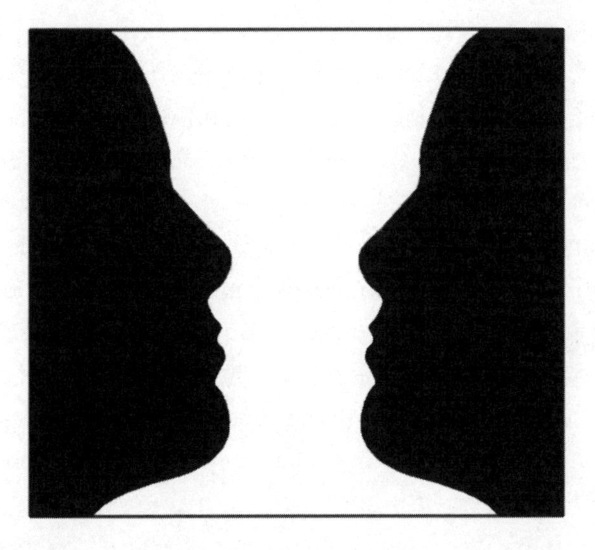

One reason it's so hard to see nothing, in other words, is simply that it only becomes nothing when you don't pay attention to it; and then, you have no way of knowing that it's there.

SEEING NOTHING II: MAKING SOMETHING OUT OF NOTHING

Humans routinely drive 100 *miles per hour*
on the autobahn, perform brain surgery, and juggle
flaming torches, even though a portion of what
we see is merely a good guess.
—LEON LEDERMAN, *The God Particle*

How do you see a quark? Not directly, that's for sure. Physicists piece together bits of tracks left in electronic detectors by fleeting subatomic particles. From these woefully incomplete clues, they deduce the presence of quarks, mesons, and the rest. If this seems unfair—too indirect to be considered "real seeing"—it is no different than the way the eye and brain create entire visual worlds out of meager, splotchy, and distorted information. Most of what we see is created in our heads.

Consider a few examples. The back of the eye contains a kind of screen, the retina, which collects light that has entered the pupil and been focused through the cornea and the lens. But there is a big hole in this screen. Each eye has a substantial blind spot where no information is recorded—the place where the optic nerves leave the eye and head for the brain. These blind spots are black holes for information coming into our heads from the outside world.

And yet we do not see "holes" in our field of vision. Our brains simply extrapolate from the surrounding scene into the void, filling in what they think should be there. Remarkably, this process usually works. Even more remarkably, most of perception is based on a similar process of filling in holes, making guesses, jumping to conclusions. We judge every book, so to speak, by its cover. A deft caricature can evoke a whole personality. A glimpse of moving red light translates into the make, model, and speed of a receding car.

This ability to make something from nothing, while useful, also creates a lot of somethings that aren't really there: the man in the moon, the dragon in the cloud, the monster in the bathroom tiles. It leads people to see qualities in others that aren't necessarily there as well, projecting fears, hopes, expectations. Sometimes, it leads scientists to see results in experiments that are more a creation of wishful thinking than objective analysis of data.

In fact, it may be even easier to make something of nothing than nothing of something. Perceptual psychologist Ann Treisman, of UC Berkeley, found that people shown a field of identical letter Q's with a simple O hidden in the middle did not see the O. The brain probably added the "tail" necessary to turn the O into a

Q. And yet, people shown a field of O's had no problem finding a solitary Q.

Finding the presence, in other words, was easy; but finding an absence was impossible for most people—even though the two situations were in all respects mirror images, entirely complementary. The lack disappears right down the proverbial void.

Other senses are affected as well. Some people "hear" voices, sounds, ringing in the ears—even in the complete absence of external speakers or bells. Several readers of Martin Gardner wrote to tell him about the "Bowery El phenomenon": after the Third Avenue El was torn down in New York, people woke up in the night hearing strange noises. They called the police—only to find out that what they were sensing was the absence of the rattling of trains they had become so accustomed to hearing.

Other people are affected by phantom tastes and smells. Almost always associated with nerve damage, these sometimes overpowering sensations of rotten tastes and awful odors have driven some patients to consider suicide, according to Linda Bartoshuk of Yale University. But there is little in the medical literature quite as strange as the studies of phantom limbs conducted by neuroscientists such as V. S. Ramachandran and Oliver Sacks; while the former studied people who had lost limbs, yet felt them vividly, the latter, as we shall see, watched his leg evaporate into "nothing" after an unfortunate encounter with a bull.

FEELING NOTHING

> *There is a deeper message here:* Your own body *is a phantom, one that your brain has temporarily constructed purely for convenience.*
>
> —V. S. RAMACHANDRAN, *Phantoms in the Brain*

The experiences of Sacks and Ramachandran are complementary: Sacks sensed something as nothing; Ramachandran's patients felt nothing as something. Both lead us to the same unsettling conclusion: the body, like the ether, is a model that the brain constructs. Usually it works. When it ceases to offer the appropriate, consistent, information, the brain—like physics—changes the model.

Ramachandran's research was spurred by his frustration over watching helplessly as many patients suffered a variety of symptoms—including severe pain—in limbs that no longer existed. "Chronic pain in a real body part such as the joint aches of arthritis or lower backache is difficult enough to treat," he writes, "but how do you treat pain in a nonexistent limb?"

His research (and that of others) has shown that the ever-elastic brain maps each part of the body onto a specific area in the cerebral cortex. When that part of the body is damaged and no longer sends signals to the cortex, the brain reconfigures its map; in effect, it fills in the missing sensations by rerouting nerve impulses from another body part. For example, the brain may transfer sensations previously coming from a now-missing hand to a cheek. Thus, a patient who has lost a hand may still feel a complete range of feelings in that (no longer existing) hand—but only when someone touches his cheek. One patient Ramachandran describes experienced sexual pleasure in a nonexistent foot. When one body part was unavailable to provide sensation, the brain simply borrowed another and mapped those feelings onto the nonexistent "phantom."

What, then, is the body? Real or phantom? Something or nothing? It is one thing to consider the something and nothing of physicists as purely arbitrary distinctions, constructions subject to continual revision as knowledge and circumstances change. It is quite another to apply the same analysis to the knowledge we gain from our senses: the everyday things we see and hear. Most unsettling of all is to question the very concrete "somethingness" of our own physical bodies.

"For your entire life, you've been walking around assuming that your 'self' is anchored to a single body that remains stable and permanent at least until death," writes Ramachandran. "Yet these experiments suggest the exact opposite—that your body image . . . is merely a shell that you've temporarily created for successfully passing on your genes to your offspring."

Oliver Sacks became a victim of a syndrome that was oddly the reverse of that experienced by Ramachandran's patients. As he recounts in *A Leg to Stand On,* the neurologist stumbled hard and severely injured his leg while running away from a bull that he surprised (and vice versa) during a hike. For an extended period of time, his leg was immobilized in a cast; he had no feeling in the limb; even what he could touch seemed unreal, not his, a disembodied thing that someone had somehow stuck on, an impostor.

He had, as he puts it, "lost my leg. It had vanished; it had gone; it had been cut off at the top. I was now an amputee." Except that the leg was still, inexplicably, there. Sacks became, in his own words, an "internal" amputee. He had terrifying dreams that the cast that held his leg was actually hollow, empty, "a chalky envelope, a mere shell, a nothingness, a void . . . a leg impossibly made of nothing whatever."

Such feelings are common, Sacks writes, in patients who are given anesthesia that numbs the lower half of the body. Those patients do not, in general, feel numb in those nether parts. Rather, he says, "one feels that . . . one has been cut in half, and that the lower half is absolutely missing—not in the familiar sense of being somewhere, elsewhere, but in the uncanny sense of *not-being,* or being nowhere."

It's worse than that: Not only does the body cease to exist; the space in which the body lives does not exist. That whole part of the world, and everything in it, evaporates completely. Sacks refers to this as a "hole" in consciousness. Just as amnesia is a "hole" in memory, so a person can experience a "hole" in spatial awareness.

While recovering from his accident, Sacks chatted with a fellow patient—an amputee who was tortured by pain in his no-longer-existing "phantom" leg. "Isn't that the darnedest thing!" said the companion. "Doc here's got a leg, but no feeling in the leg—and I've got the feeling, but no leg to go with it! You know, we could make one good leg between us."

Which is something, which nothing? The "real" leg with no feelings attached? Or the "real" feelings with no leg attached?

In perception, at least, an absence is every bit as real as a presence. The absence of gravity makes astronauts as queasy as any stomach flu. An absent limb can cause as much pain as a real, physical, one. To some extent, absence and presence in perception are entirely interchangeable. How, then, does the mind come to conclusions about what is something and what is not? In some ways, the methods it uses are highly reminiscent of those employed by physicists and mathematicians.

RENORMALIZING NOTHING

When you drive down the road and the radio is blaring and the wind is blasting past the windows, the radio doesn't sound loud, because you move your zero. You're shifting zeros all the time.

—THOMAS HUMPHREY,
artist and senior scientist, the Exploratorium

What does it mean to perceive zero sound, zero light, zero temperature, zero motion? Certainly, we do not perceive zero temperature as an absence of temperature: absolute zero (–459 degrees Fahrenheit) would certainly feel cold. In terms of temperature, "nothing" would be more like room temperature—the normal everyday "lack" of temperature we take for granted. Zero temperature is temperature that you don't notice, that doesn't in any way impress itself upon your consciousness. The physical zero of temperature falls in the wrong place for the human senses. Perception moves it to the "right" one.

In fact, perception of temperature, like other sense perceptions, is almost entirely relative—a perception not of absolute quantities, but of changes. A striking illusion proves the point. Rest one hand in cold water, the other hand in hot water. Then plunge both hands into lukewarm water. The hand that was cold now feels quite hot; the hand that was hot feels quite cold. In absolute terms, both hands are experiencing the same temperature. But they perceive only the difference, the change.

The same is true of all the senses. A dark shadow under a tree on a sunny day can be much brighter in absolute terms than a white wall inside a dark room. And yet, people see the former as black, the latter as white. If you put them side by side, the dark shadow would look white, the white wall black. In the context of the bright day, the shadow looks black; in the context of the dark room, the wall looks white.

Or consider the full Moon, a bright object if ever there was one. What color is the moon? Rocks brought back from the moon by astronauts are very dark gray. In fact, the moon reflects only about 8 percent of the sunlight falling on it. It only looks bright in comparison to the black sky. The zero point of brightness perception, like that of temperature perception, is not zero light, but rather a medium gray. A totally dark room appears to the senses not as black, but as a murky gray color. If you then introduce light, the areas surrounding the light look black. If you surround the dark moon with black, it looks bright.

Perception, in other words, is all about deviation from the norm. If the temperature is colder than the norm, or hotter, we experience it as cold or hot; if light is brighter than the norm, we see brightness; if darker, darkness. If sound is fainter than the norm, we may experience its complement (as in the Bowery El experience). In effect, we shift our zeros to the middle ground. We calibrate "nothing" anew every time we enter a new perceptual world: When the rushing of the wind by your ears in a fast car becomes a "nothing" background, for example, then the radio blaring over it sounds only normally loud. When you stare at a waterfall then suddenly shift your gaze to the solid ground, the ground may briefly appear to move in the opposite direction. Your brain has temporarily recalibrated the steady motion of the falling water as no motion, or zero. When you look at the ground, it appears to be moving in the opposite direction from the waterfall.*

And so it goes. Stare at a bright red image. Then shift your gaze to a white wall. The "after image" that floats in your view will be green, or white minus red. Spin around to the left. Stop. The room will appear to spin to the right. The brain renormalizes the zero point and shifts the entire scale one way or the other. You don't notice the tight shoe until you take it off, just as you don't notice the taste of the tap water in your own city.

"The perception of nothing really means nothing changes," says Richard Brown. "Nothing means that the *derivative* [the rate of change] is zero."

In a sense, we renormalize the starting point for perception in the same way that physicists renormalize the vacuum energy by setting it to zero. The zero is arbitrary. It's also irrelevant. If there's nothing to compare it to, it could be anything at all.

---

*This renormalization only works within certain limits, of course. Beyond a certain threshold, we don't adapt: loud is loud, hot is hot, red is red.

THE MEASURE OF NOTHING

The most dramatic evidence of understanding is seeing significance in nothing. This is the point of experimental controls, and a great deal comes from nothing happening in null experiments. Only when the situation is understood conceptually is it possible to appreciate nothing. So "seeing nothing" is a strong sign of understanding.

—RICHARD GREGORY, "Science through Play," in *Science Today: Problem or Crisis?*

Clearly, Nothing can elude even what we like to call "direct" perception as easily as it slips away from the more calculated advances of mathematicians and physicists. Very often, what at first appears to be "nothing" becomes very much something if you only tweak it in the right way, ask it the right questions.

Take a blank white surface—containing no information. Cast a shadow with your hand, and allow only a small bit of light to seep through. If the surface is in the sun, you will see a small image of the sun on the paper. If the surface is indoors, under artificial light, you will see a small "pinhole" image of whatever light source is illuminating the surface. As artist Bob Miller likes to point out, the entire surface is covered with images of the sun; you just can't see them until you "subtract" all but one.

Or take a beam of white light—containing no color. Pass it through a prism. Presto, color. All you've done is separate the colors that were already there. White is simply the eye's renormalization of color to "nothing." In fact, white light may contain all the colors; it's not nothing, but everything.

Or take a clear blue empty sky. The sky is full of unseen stars. Why do they appear as "nothing"? In absolute terms, a star in the daytime sky adds as much light to the background as a star at night. The star hasn't changed, but the nothing surrounding it has.

A "pinhole" image of a solar eclipse.

In fact, the empty blue sky that we see as nothing so overwhelms the stars that they completely disappear.

Getting rid of these noisy all-pervasive nothings—called "backgrounds"—is a major preoccupation of science. There are physical backgrounds, perceptual backgrounds, even statistical backgrounds. Say you are trying to decide whether your "lucky" sweater helps you get a parking spot at work. You wear the sweater for ten days, and during five of those days, you find a spot. You could easily con-

A pinhole "shadow" is the negative, or complement, of the pinhole image. The object casting the shadow can be any size so long as the distances between it, the light source, and the surface are sufficiently large.

clude from this experiment that your sweater worked half the time. However, this would ignore the role of random chance. Assume the random chance of finding a spot was also 50 percent. Then the 50 percent "success rate" for your sweater evaporates into background as readily as the stars disappear into the daytime sky. Backgrounds, like focus, can change something into nothing, and vice versa.

Measuring nothing is hard, but it's also necessary for the accurate measurement of many things. You need to know where zero sits in order to place quantities above or below it. Even when nothing's invisible to our senses—to our best theoretical and experimental tests—it has a tangible effect.

That's why chemists, for example, must pay such careful attention to the solvents (the background) in which reactions occur. And why physicists respect nothing quite so much as an honest, clean, null result. It takes a special kind of courage, Canadian physicist Erich Vogt told his fellow experimentalists in an after-dinner talk, "to continue to push limits down and find nothing. Each one of you has normally to go home to a family after a long day's work and report courageously, 'Nothing again today.'"

And yet, he told them, nothing is more important than experiments that turn up nothing at all. "A nobody is bound to find something. It takes a somebody to find nothing."

There are, as always, social and political aspects to seeing nothing as well. Pythagoras in the sixth century B.C. found it perfectly natural to count slaves as "nothing," according to Eric Temple Bell in *The Magic of Numbers.* Slaves, like machines today, were simply taken for granted. These days, we take for granted everything from homeless people sleeping in the street to telephones and computers. We have learned to renormalize these things as part of "nothing." Whatever is standard becomes effectively invisible. For a long time, the absence of women and minority groups from science was considered so normal that no one—except a few ornery women—even noticed.

Only when something happens to nothing do we finally take note. The rustling of the leaves reminds us of the unseen wind; the orbits of planets remind us of the warped geometry of spacetime; gravity feels the weight of the vacuum even when we cannot.

ZEN ZERO

The only possible access we may have to phenomena that transcend human concept and sensory perception is by cultivating states of awareness that themselves transcend language, concepts, and sensory experience.

—B. ALAN WALLACE,
Choosing Reality: A Buddhist View of Physics and the Mind

We shape clay into a pot, but it is the emptiness inside that holds what we want.

—TAO-TE CHING

Perhaps no discipline has taken the perception of nothing so seriously as Buddhism, particularly Zen. In Zen, the understanding and attainment of nothing is cultivated and celebrated as the path to both spiritual peace and deep understanding. This nothing has nothing to do with nihilism. The Buddhist void, like the quantum vacuum, overflows with potential. It is the empty gourd that contains everything. It is a void shimmering with possibilities and endless potential. It is not empty; it is waiting.

In fact, without emptiness, neither understanding nor spiritual growth are possible, according to Buddhist teachings. Emptiness and fullness are not opposites, but part of the same larger reality; they are two aspects of the same thing, each necessary for the existence of the other. Buddhists cultivate emptiness because only nothing, they say, can make room for something. If the cup already overflows, you can't put anything inside. A blank mind invites insights that simply can't be heard over the normal everyday noise generated by the normal everyday brain. Zen practice clears the cobwebs from the mind itself; like cleaning one's eyeglasses, it erases the multitude of distractions that prevent us from seeing what is right in front of our noses. Only a mind full of nothing can be open to everything.

Psychology has also tested these waters, and found them welcoming. New York University psychiatrist Mark Epstein, for example, incorporates meditation into his practice. He concludes that the mind can be at rest only when it learns to tolerate nothingness. "Emptiness is vast and astonishing, the Buddhist approach insists," he writes in his book, *Going to Pieces without Falling Apart.* "It does not have to be toxic. In the end, he concludes, "we can know ourselves only by surrendering into the void."

While Zen teachings certainly do not share the well-tested validity of scientific knowledge, in recent years, there has been increasing conversation between Buddhist scholars and physicists—primarily on the places where the emptiness of mind and the emptiness of vacuum overlap. UC Santa Barbara Buddhist scholar Alan Wallace, who was trained in physics at Amherst College, works with the Dalai Lama and top physics researchers throughout the world to explore these rich and curious intersections. Many have to do with the obvious interest of both groups in properties of the quantum vacuum—the roiling physical nothing that gives rise to all things.

However, an even more interesting commonality concerns the nature of reality itself. In both physics and Buddhism, every coin has multiple, often mutually contradictory, sides. Particles and fields and forces are to some extent nothing but models made up in the mind's eye. And yet, you cannot walk through a wall made of particles and forces. Walls, like phantom limbs and zero and the funny energy that pushes the stars around, are both real and unreal—something and nothing.

Reality, as both Buddhism and quantum physics tell us over and again, requires two inextricably interwoven players: the observer and the observed. Physics tends to focus on what is observed. That requires various mental and mathematical models that help pin vague concepts in their place, make them amenable to precise manipulation. Only with the models in place is it possible to ask concrete questions.

Zen practice, however, tends to the observer. Every experiment, no matter how carefully prepared and monitored, begins and ends with the human mind. This is the ultimate instrument of science. There is no contradiction here. The fact that the vacuum is just a convention "by no means implies that its existence is arbitrary," Wallace argues. "The laws of physics . . . are precisely determined by means of experiment and observation. They are not simply creations of arbitrary human whim. They have no independent existence, however."

While many would disagree with his conclusions, it is hard not to see some sense in the implications of these teachings:

In the end, one way to see clearly through both the noise from the outside world and the perhaps even noisier rumblings of our own restless senses is to quiet—or at least better understand—the mind itself.

Chapter 10

IN SEARCH OF NOTHING

That "nothing" from which something arose should not . . . be confused with the emptiness of a vacuum. It is nothing in a profounder sense. It is nothingness.

—JOHN WHEELER, *Geons, Black Holes, and Quantum Foam*

SO WHAT IS NOTHING, anyway? Not the vacuum or the empty stage of spacetime on which the drama of the universe plays out; not zero or null results; not silence or darkness; not blank pages or black holes. The entirety of this book has been devoted to demonstrating that most conventional notions of nothing are most certainly something; they do not support nothing's reputation as a big zero.

However, the concept of a "true" nothing—like true love—still calls to us from the void. If it's there, we'd like to look it in the face.

More important, this ideal nothing remains an essential missing piece in the puzzle of the universe. Until it's understood, the search for it will likely continue to play a critical role in our thoughts, in mathematics, in science.

Is there such a thing as pure, undiluted nothing? Can it be distilled? If so, what might its properties be?

Clearly, human notions of nothing have changed over time and will continue to change. However, some ideas seem to stick around while others fade into the forgotten backdrop. Why was the ether an error when the modern vacuum persists, even as it gets more and more complicated over time, done up like a Christmas tree with the cosmological constant, quintessence, Higgs fields, and heaven knows what?

One reason is certainly that the vacuum was allowed to evolve, while the ether was relegated to a role that quickly became irrelevant and outdated. At least in the minds of physicists, the term "vacuum" has expanded to embrace every newfangled accoutrement that's come along. Like "particles" that evolved from points to waves, or atoms that made the leap from solar systems to resonating bundles of vibrations, the vacuum has accommodated to the modern age.

"Why did the electron survive, while the ether did not?" asks Alan Wallace. "The simple answer is that physicists simply decided that the meaning of the term *ether* would be frozen in its classical context, whereas *electron* was allowed to adapt to a sequence of changing theories."

In the same way, space and time have become radically different species since they were first bandied about in the minds of ancient thinkers millennia ago.

And what is to become of nothing? Only time and physicists will tell. One thing is certain: the old explanations no longer suffice. The new nothing will have to learn to live with roiling vacuums and warping spacetimes, multiple dimensions and repulsive forces. It's tempting to conclude that there's nothing, in the end, that's really *nothing.* Or that nothing is merely a convenient stage upon which to study the various somethings that concern us.

But the history of science suggests that nothing is not an easily ignored nonentity. However elusive nothing might be, it remains the starting point for everything. And until nothing is understood, nothing can be.

What follows is not so much a series of conclusions about the nature of nothing as a survey of some seminal ideas that will almost certainly play a central role in the evolving understanding of emptiness. And, of course, a happy ending.

A MESS OF NOTHING

{Something} offers resistance to touch, however light and slight, it will increase the mass of body by such amount, great or small, as it may amount to, and will rank with it. If, on the other hand, it is intangible, so that it offers no resistance whatever to anything passing through it, then it will be that empty space which we call vacuity.

—LUCRETIUS, *De Rerum Natura*

What is the difference between something and nothing? There are many answers, some more useful than others. If you argue, as Lucretius did, that something is what pushes back, you risk relegating a lot of things to nothing that surely are not; color doesn't push back, nor laughter, nor even—usually—air.

Nothing is not that simple.

A better answer might be: Nothing is a mess. Something has some sense of order, of form. Nothing has no discernible properties, no pattern, no structure. Humans, at least, perceive things by noticing patterns. If nothing has no pattern, it would be imperceptible—which is to say, not there—at least to us. It would be so disordered that we wouldn't know we were standing inside it, like a cloud, or fog.

The closest anything in our universe comes to this kind of nothingness is probably a "melted" vacuum—a nothing without even the minimum structure that gives emptiness its shape, without even the minimal properties that create the potential for forming the particles that populate the universe.

Of course, this melted vacuum—by definition—has energy. It takes energy to melt the vacuum, to destroy the innate order that creates our universe, so energy has to be a part of that particular kind of nothing.

Can you take away energy, too? Yes, but when you reduce energy to its bare minimum, at least in our universe, you get a structured, "frozen," vacuum. In our universe, it appears that you can have one kind of nothing or the other, but not both at the same time. That is, you can imagine a structured vacuum like the one we live in, the lowest possible (so far as we know) energy state. Structure, yes; energy, no. Conversely, you can imagine a very disordered, chaotic, hot state such as existed at the very beginning of time—the melted vacuum. Energy, yes; structure, no.

Of course, to get to true nothing, you'd also have to take away space and time. The vacuum is one kind of nothing. The background of space and time is another. These two kinds of nothing are related, but not the same. "Could you have a vacuum without spacetime?" pondered CERN physicist John March-Russell. He figured you could. A single atom in its minimum energy state, for example, could exist in its own speck of vacuum without space and time. But could you have space and time without a vacuum? March-Russell didn't think so.

"The current belief is that you have to understand all the properties of the vacuum before you can understand anything else."

Assume that you could eliminate both the vacuum, and space and time. That leaves only the laws of nature themselves. But the laws of nature are orderly; their regularity is what makes them recognizable. True nothing would be an absolutely lawless realm. And physics doesn't go there. At least not yet. (If it did, it wouldn't be physics.)

There is one way to have a kind of "pattern" that is completely imperceptible. That "pattern" is perfect symmetry—and I put the word *pattern* in quotes because, as we shall see, truly perfect symmetry contains no discernible pattern at all.

Symmetry, in a sense, is immunity to change. If you can transform something such that the transformation doesn't make a noticeable difference, that's symmetry. Mirror images fool people because they can't tell an object from its image. But mirror images are very limited kinds of symmetries. For one thing, they reverse left and right, so the symmetry isn't perfect.

If something were *perfectly* symmetrical, then no matter how you tried to change it, the hypothetical change would have no effect. Without change, there is no perception. A perfectly symmetrical nothing would be a state so changeless that nothing you could conceivably do to it would make a difference. (We'll come back to this idea in a bit.)

Alas, our universe is not perfectly symmetrical. Exploring the properties of nothing has to take place—at least primarily—in the universe we know and love. And if you take away everything you can possibly take away from a patch of our present universe, you still have structure. Finding out the nature of that structure is a major preoccupation of contemporary physics.

The search for nothing, then, starts with the vacuum.

WELL-ORDERED NOTHING

Once we find the right way of describing it,
the vacuum is nothing.

—HELEN QUINN

What is crucial is to find just one vacuum,
empty, and stable. . . .

—MARTIN H. KRIEGER, *Doing Physics*

The nothing most amenable to physics—at least for now—is the vacuum, or vacua, of empty space. That vacuum is, by definition, the lowest possible energy state—the state you get when you take everything else away. Think of it as the absolutely calm surface of water, or a sheet of paper with nothing on it, or an endless stretch of absolute quiet. There's something there, to be sure, but there's nothing going on in it.

So far as physicists know, there are two things left in the vacuum even when you take away everything that's possible to take away. One is the fabric of spacetime. The other is the Higgs field—the structure of the vacuum that appears to give particles mass.

Although the Higgs field is still a theoretical entity, it qualifies as an irremovable part of the vacuum in current theories because it appears to be part of the universe at its lowest possible energy state. To get rid of the Higgs field requires going to higher energy—a false vacuum (see Chapter 4). The lowest-energy (therefore "authentic") vacuum seems to have the Higgs field. The Higgs field also qualifies as nothing because we can't do anything to it that tells us that we're in it; the Higgs field to our universe is like water to a fish—the same everywhere and therefore utterly imperceptible. "Whatever the [modern] ether is," says Jan Rafelski, "as long as it is invariant, everywhere the same, and we have no means to pump or just move it, it is effectively invisible, it is the 'nothing' of our classical physics."

One of the biggest criticisms of string theory, as we have seen, is that it can't produce a good workable nothing. The theory has no single vacuum or ground state; it has tens of thousands of families of ground states (see Chapter 6). With all these different possible vacua floating around, it's difficult to get a handle on the theory or the kind of universe(s) it might produce. It certainly doesn't (as yet) appear to produce the kind of vacuum that leads to a universe like ours.

Each of the vacua in string theory corresponds to a different configuration of the extra, hidden, dimensions; depending on how these strange extradimensional landscapes weave together and unfurl, a completely different set of physical laws emerges, with different possibilities for particles, forces, and everything made of them. This embarrassment of riches is part of what makes it almost impossible to link the beautiful mathematics of string theory to the real universe.

String theory's biggest problem, in other words, is that it can't find the right nothing.

Whether or not the universe turns out to be stringy, fields provide an underlying structure that connects everything to everything else. Space, time, matter, and force are connected like lines between dots; if the dots turn out to be unimportant (as, indeed, they already have), then what's left is the structure of the web itself, the pattern of interrelationships.

Everything in the universe, then, is at heart a pattern of relationships. The question remains: What determines which patterns are possible? One answer is nothing.

THE POTENTIAL FOR NOTHING

Real systems are, in this sense, "excitations of the vacuum"—much as surface waves in a pond are excitations of the pond's water. . . . The vacuum in itself is shapeless, but it may assume specific shapes. In doing so, it becomes a physical reality, a "real world."

—HENNING GENZ, *Nothingness: The Science of Empty Space*

You cannot ice-skate on a soccer field or play golf on a tennis court. The shape, size and texture of a playing field limits the kinds of games you can play on it. In the same way, the vacuum determines what kinds of phenomena can sprout from its not-so-empty depths.

In this sense, nothing is the potential to make things happen. Depending on the kind of nothing in place, different somethings can emerge. As we have seen, a false vacuum—even though it is "nothing"—can create the universe because it has potential, just as a brick on the roof of a building has the potential to fall and hit someone in the head. All it takes is a nudge in the right direction.

When physicists talk about the structure of the vacuum, they are really talking about the kinds of things that can be produced by and in a particular vacuum. "The structure is just another way of thinking about the things that could be in it," said Lenny Susskind. "It has the potential to have things, and different vacuums have different potentials for different things."

In fact, the quantum vacuum seems to require that something emerge from nothing. Because nothing is impossible in the quantum vacuum (and—most important—"nothing" itself is impossible) the question of why the universe is here is answered by the existence of quantum mechanics itself: In a quantum mechanical universe, some kind of universe *has* to be here. The only thing we don't know is Why quantum mechanics? Why laws of nature at all?

"If this approach is right," writes Alan Guth, "perpetual 'nothing' is impossible. If the creation of the universe can be described as a quantum process, we would be left with one deep mystery of existence: What is it that determined the laws of physics?"

In addition to the potential inherent in the vacuum, there is also a potential built into the nothingness of space and time. Space and time, after all, are the stuff of gravity. Gravity is the curvature of space and time. So the potential locked in the brick sitting on the roof is, in a very real sense, stored in the warpage of spacetime itself. This potential energy of spacetime, as we have seen, powered the explosive growth of the universe, the creation of matter, and everything that followed.

Curiously, this notion of nothing as potential reflects a very Buddhist perspective: nothing as infinite potential, or what Zen teachers call "the inexhaustible storehouse with not a single thing in it." The emptiness of nothing creates the possibility for infinite kinds of somethings.

Nothing is the potential for everything.

What happens if space and time dissolve, however? Does the potential for anything disappear? Does the dissolution of space and time mark the true beginning of nothing? Both string theory and cosmology are providing hints that this indeed might be the case.

NOT EVEN SPACE AND TIME

The history of physics is the history of giving up cherished ideas. Space and time are things we've cherished for thousands of years, and it's clear we're going to have to give them up.
—ANDREW STROMINGER

For now, strings are simply inserted into a background of space and time. However, as discussed in Chapter 6, string theorists understand that this is not the ultimate solution. After all, some string theories without space and time seem to be equivalent to other string theories with space and time. If you can get rid of space and time that easily, clearly they are not fundamental properties. That means string theorists will have to find a way to create string theory out of whole cloth—without space and time as a backdrop. They will have to find the blank slate that brings space and time into existence.

Space and time certainly appear to have different properties at the scale of strings—this much is already known. Specifically, as we have seen (Chapter 6), geometry becomes "noncommutative"—that is, $x \times y$ no longer equals $y \times x$. The mathematical underpinnings of space appear to be changing. It isn't clear what will take their place. Whatever the outcome, it's clear that physicists—or at least, string theorists—are prepared for a major revolution in the way we think about space and time. "Many people believe that there will be a major dismantlement (of space and time)," said

Greene. "We will not understand string theory until we make a major break with notions of space and time."

In fact, much of the discussion at a recent meeting on string theory at Caltech concerned these very questions. Two thousand years ago, David Gross reminded his colleagues, Democritus argued that qualities such as "sweet" or "bitter" were merely conventions. In reality, there were only atoms. Today, he said, the conventions we have become accustomed to are everyday space and time. In reality, we don't yet know *what* there is.

As for cosmologists, their task is to dig back into cosmic history for clues as to how space and time came into existence in the first place (see Chapter 7). Whatever the outcome of their efforts, it's almost certain that space and time as we know them melt into something unknown—and perhaps unknowable—at the very beginning of the universe.

NOTHING FOR ME

If the universe hadn't been suitable for our existence,
we wouldn't be asking why it is the way it is.
—Stephen Hawking

Sweet nothings, to whom shall they be murmured?
When there is no one to hear them, and no air
to vibrate to carry the sound?
—Roald Hoffmann

Why isn't there nothing? Why are we here to ask the question? At least one line of thinking suggests that the human presence, of all things, plays a role in the answer.

According to this view, the fact that the universe is the way it is necessarily reflects the fact that we are here to study it. Either that, or the universe is a very lucky coincidence. Consider: If electrons weighed a little more and protons a little less, if gravity were a little stronger and the nuclear force a little weaker, atoms and stars wouldn't exist. Without atoms or stars, nothing familiar—including life forms—could exist either.

The same is true of the structure of the vacuum, or the existence of space and time. Perhaps there are billions of possible vacua, billions of possible configurations of space and time, billions of possible geometries for hypothetical extra dimensions. Only one, however, is amenable to life. If so, the all-important question of why we happen to inhabit the one we do may not be a matter of logic or natural law. It may simply be the only universe that's hospitable and therefore the only universe we can know.

If this is true, then the nature of something may boil down to us, after all. If nothing is what we can't perceive, then something is, by definition, what we can perceive. We perceive the only universe we *can* perceive—so in a sense, our very perception determines the kind of universe we populate. This self-centered view of something and nothing seems appropriate for the so-called Me Generation. Physicists prefer to call it the anthropic (or human-centered) view.

Most physicists actually abhor the idea that humans could influence the very definition or structure of the universe. Physicists like the laws of nature to pop out of pure logic, and the universe to emerge, well, naturally. Whether or not life in any form exists should have nothing to do with the way the universe begins or evolves.

It "would be a sad outcome," if that were not true, said Greene. It would rob physicists of their motivation to find the ultimate "why" behind the laws of nature, he said. The answer to "why" would simply be an accident of evolution: physicists exist.

On the other hand, it's very difficult to ask questions of nature that aren't somehow already colored by our very human preconceptions. Even the simplest, most objective, questions may play into preexisting prejudices.

"You think you're asking a neutral question, but there are things you believe that you don't know you believe," says Sidney Coleman. "They are too deeply embedded in the language."

These questions are far too big to be answered directly in any event. Science rarely proceeds by asking enormous (and enormously vague) questions like What is nothing? Or even, What is empty space? Or, Why are we here? The way to explore the forest is tree by tree. A more manageable way to get a handle on nothing is to consider its mathematical properties.

NOTHING'S PERFECT

Zero means don't change. Don't move. Zero is the one that doesn't make a difference. That's an honest-to-god Zero.

—DAVID HOFFMAN, Mathematical Sciences Research Institute, Berkeley

But the most surprising and wonderful composition was that of whiteness. There is no one sort of rays which alone can exhibit this.

—ISAAC NEWTON, *The New Theory about Light and Colors,* communicated to the Royal Society, February 6, 1671

Nothing doesn't register. Nothing doesn't make a difference. Nothing poses no obstacles and adds nothing to the equation. Nothing has no features, and nothing can't change. Nothing has no direction, no size, no preferences.

Nothing, a mathematician might say, is perfectly symmetrical. To most people, symmetry describes something like a snowflake, or a border of bathroom tiles, or maybe even a Rorschach inkblot. There is a certain sameness between one part of the object and another part. One can be superimposed on the other—perfectly. A snowflake can be rotated through 60 degrees, and you won't be able to tell that you rotated the snowflake. Another one of the snowflake's six points may be facing up, but since all points are the same, you can't tell the difference.

Perfect symmetry is even more immune to change. A circle is more symmetrical than a snowflake because you can rotate it through 5 degrees or 10 or 17, and it will still look exactly the same. A sphere is even more symmetrical than a circle because you can turn it in three dimensions and no matter how you turn it, it looks the same. Truly perfect symmetry is impossible to imagine because it would be symmetrical in every possible respect. Things would have to stay exactly the same as they moved forward and backward in time, for example. They would have to be able to change shape, change strength, change motion, change size, change any quality or quantity you can think of and still stay exactly the same.

The closest we can probably come to imagining perfect symmetry is a smooth, timeless, featureless empty space—the proverbial blank slate, the utter silence. It can't be perceived because nothing can change. Everything would be one and the same; everything would be the same, as far as we could tell, as nothing.

If nothing is perfect, however, then something is nothing shattered, symmetry broken. It is altered, imperfect, nothing. A white piece of paper has "no" color, no information. Break the symmetry by subtracting some color (with a filter or prism or pen) and suddenly there is something where there was nothing before. Newton himself was amazed that the color white was not a single color, but an unbroken mixing of every visible wavelength in the palette of the Sun.

Most physicists believe the universe we live in is just such shattered perfection. If nature were understood at the deepest level, they have reason to think, it would be far more symmetrical than anything that exists in the universe today. Leon Lederman, in *The God Particle,* compares this primordial state of perfect symmetry to the biblical town of Babel before the people tried to build a tower all the way to heaven. Everyone spoke the same language, so in that sense, the situation was very symmetrical. When God got wind of their efforts, however, he got angry. To stop the tower project, he made everyone speak a different language, so that no one could communicate (and therefore cooperate) with anyone else.

God, in effect, broke the symmetry of Babel.

In the same way, our universe is a kind of crippled nothingness, a series of broken symmetries. This is a good thing. If the symmetry between matter and antimatter were perfect, for example, then all the matter and antimatter would have annihilated and there would be no matter left. If all the particles of matter were perfectly symmetrical—that is, if electrons were indistinguishable from protons—there would be no atoms, or us. If all of the forces that propel the particles were symmetrical—if gravity were the same as electromagnetism—then there would be no stars, or starlight. If time were perfectly symmetrical, there would be no past or future.

Almost everything we know and value is the result of broken symmetry. An undifferentiated fertilized egg breaks symmetry many times as it evolves into a full-grown baby. The symmetry of a blank canvas is broken by paint just as silence is broken by words and music.

Still, there is an element of the fall from grace here. Creation from nothing seems to involve breaking things, a blemish on perfection.

NOTHING SHATTERED

The reality we observe in our laboratories is only an imperfect reflection of a deeper and more beautiful reality, the reality of the equations that display all the symmetries of the theory.

—STEVEN WEINBERG, *Dreams of a Final Theory*

Why and how does symmetry break? It doesn't need a reason or a why. Often, it can just happen.

Consider a pencil balanced on its pointy end: It stands at the center of infinite possibility. It can tip in any direction. If there is no breeze brushing it one way or the other, no irregularity in the graphite point, it will have no particular preference. Every direction will be equal. The situation is symmetrical.

Once the pencil has fallen, however, every direction is no longer the same. Now one direction—the one the pencil has taken—is different, singled out, preferred. Symmetry is broken. Nothing is behind this shattering of perfection but random chance.

Or think of a ball spinning around a roulette wheel—Hawking's favorite example. So long as it's spinning, it has an equal chance of falling into any of the possible slots. Every time it spins, it behaves the same. However, once it's fallen into a slot, the symmetry is broken. It has thirty-seven different possibilities for a final state—not one of them equivalent to the others.

The roulette ball, like the pencil, "falls" into a lower energy state and spontaneously breaks the symmetry that existed before. In the same way, physicists believe that our universe was once in a highly symmetrical high-energy state. Gravity, electricity, nuclear forces, were all on the same footing. Particles all had the same mass (or no mass). Perhaps all eleven dimensions (if there are eleven dimensions) were the same size. Then symmetry was broken and the universe as we know it evolved.

The universe fell from symmetry for the same reason pencils fall from their points and roulette balls sooner or later fall into slots. It is the natural place to go. A pencil wants to fall for the same reason that water wants to flow downhill; downhill is a state of lower energy. And the unsymmetrical state is usually a state of lower energy. Everywhere you look, hot (high-energy) things like to spread out, randomize, run around in chaos. Cooler (lower-energy) things like to clump, crystallize, align. Water looks the same in every direction, but ice has a crystalline structure. In ice, water's symmetry is broken.

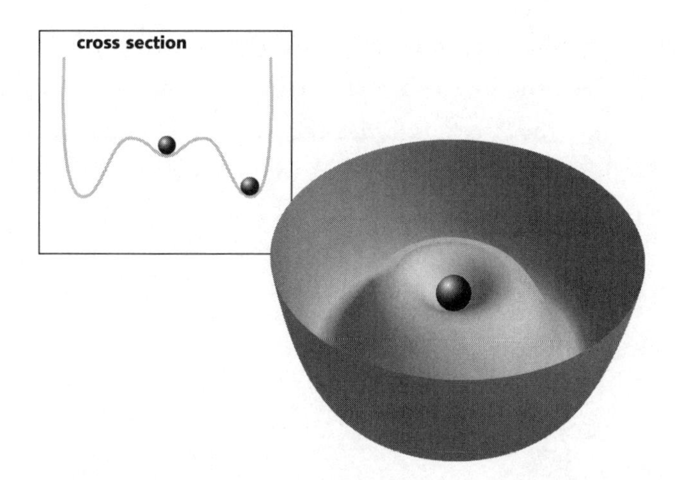

The ball could fall into a lower energy state, breaking the symmetry.

But breaking symmetry does not require a change in energy. Imagine that some kind of gas is uniformly dispersed in a room. It could be perfume or water vapor. It would be randomly distributed and therefore quite invisible. Nothing, you might say.

It is possible—though certainly not probable—that through pure chance all the perfume (or water) molecules would find themselves in the same place at the same time. If so, they would condense into a puddle on the floor. The perfect symmetry would be broken; something would appear out of nothing. And yet, no energy would be gained or lost.

Breaking symmetry makes a huge difference in how things appear—even though underneath it all they may still be the same. Consider the shadow you cast on a wall with your hand. The shadow can appear to represent many different things: a hand, a duck, an alligator; if you clench your hand into a fist, it can look like a ball; if you hold it flat, it can look like a rectangle. Looked at in its complete three-dimensional perspective, it's clear that the essential features of your hand don't change. But looked at through the narrow two-dimensional "slice" that the shadow presents, your hand appears to be a wide variety of unrelated objects. The shadow hides the symmetry inherent in your hand in the same way that freezing hides the inherent symmetry of water or falling hides the symmetry of the forces acting on a precariously balancing pencil.

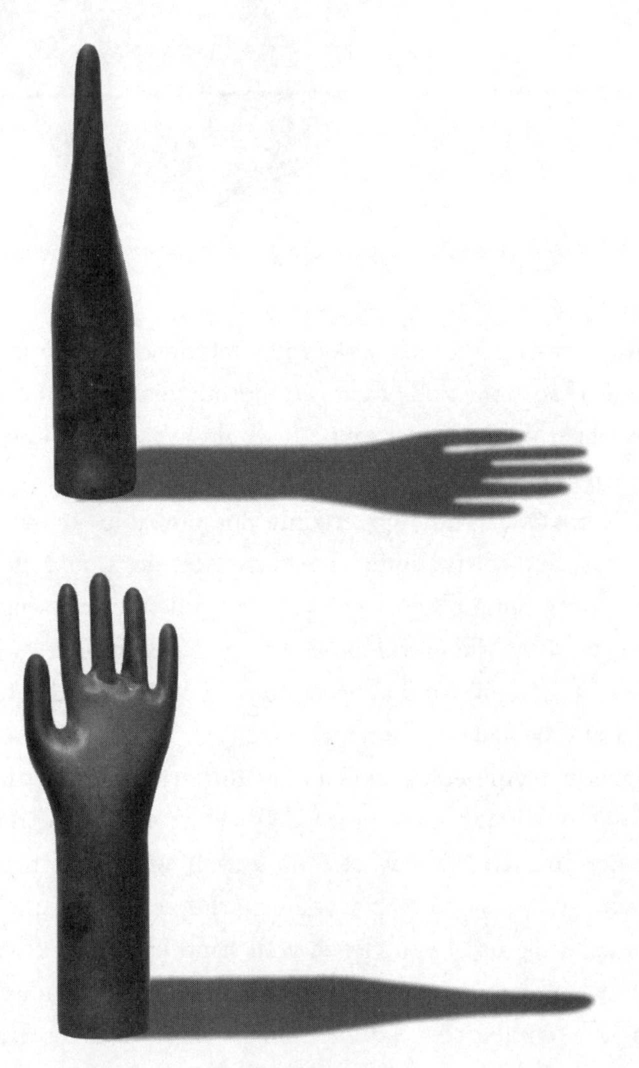

Broken Symmetry

NOTHING REGAINED

As we explore physics at higher and higher energy,
revealing its structure at shorter and shorter distances,
we discover more and more symmetry.
—DAVID GROSS, *The Role of Symmetry in Fundamental Physics*

If you could melt the frozen structure of the vacuum, you would find the perfect symmetry that existed before the universe cooled down. There is a hidden perfection in the universe that can't be seen today but can only be deduced from clues provided by particle accelerators and perhaps highly energetic cosmic events. Or so physicists believe. One reason is that the more physicists understand about the universe, the more symmetrical it looks. Electricity and magnetism were considered two completely unrelated forces as recently as 150 years ago; now we know they are only different aspects of the same phenomenon. The same is true of matter and energy, gravity and inertia, space and time.

The closer you look, the more things that appeared to be different turn out to be the same. These symmetries are hidden under normal circumstances, but clever detective work (both theoretical and experimental) can often tease them out. When and if it does, physicists may well reveal that the universe with all its various forces and particles only *seems* complicated because the underlying symmetry is no longer evident. It is as if, as Hawking put it, we can only see the ball sitting in slot number four of the roulette wheel. We don't realize that at higher energies, as the ball spins around, all the slots look the same, and the ball could equally fall into any one of them.

The farther back we dial the clock of the cosmos—either through theories, or with experiments in particle accelerators—the closer we will come to re-creating that perfectly symmetrical something (or nothing) out of which everything came. Gross likes to call this paradigm "Beauty in; garbage out." If we stay on this trajectory, perhaps symmetry will be all there is. Perfection—and nothing—regained.

The task of sorting out just how the symmetries broke will not be easy. It happened more than once. The various forces and particles grew out of different phases in the fall from perfection, different "freezings" at different temperatures, at different times along the evolutionary path. Some are already worked out; others remain elusive. However, early successes inspire confidence that the process of uncovering more symmetries will only continue—until we return to a time before the universe lost its symmetry, if not its innocence. And every time a symmetry broke, the universe narrowed its options. Like the falling pencil, the universe went from a state of infinite possibility to the reality we know today.

One question physicists would certainly like to answer is Did the symmetries have to "break" exactly the way they did to produce our universe? That is, could the universe have evolved differently? If you ran the big bang all over again, would you get gravity and electromagnetism, quarks and electrons, helium and gold, planets and life?

Or as Einstein famously put the question: Did God have a choice?

TIME WILL TELL

We must be able to answer the question
What is time? as simply and clearly as we
answer the question What is heat?

—ALEXANDER POLYAKOV, Princeton University,
String Theory at the Millennium, Caltech

One kind of asymmetry our universe can't seem to do without is time. If forward and backward in time made no difference, how would anything ever change? If nothing changed, nothing could form. And yet, in a state of pure nothingness, nothing can change because there is nothing to change. Time requires that something change—otherwise, there is no such thing as time. Things changing is what defines time: atoms vibrate; clocks tick; the earth circles the Sun; the universe expands.

Time, in other words, can't exist in nothing, and nothing can't exist in time. Time and "something" are somehow related. Everything depends on time. And nothing exists without it.

But time is a tangle of paradoxes. You can go left and right in space, but not left and right in time. Time is a one-way street. It's easy to imagine ten dimensions of space because you can just add one on top of the other. But there's only one time. And it always moves in the same direction.

Actually, there are, arguably, several arrows of time, all of which currently (perhaps necessarily) coincide. There's the familiar arrow of entropy that leads inexorably from order to disorder: eggs break, paint peels, ice melts, skin wrinkles. There's the cosmological arrow of time determined by the expansion of the universe from a point in the past to the vast expanses of today. There's the psychological arrow of time that people perceive: they remember the past, but not the future.

Why should time have so many arrows when space has none?

More confusing still, on the level of subatomic particles, time *does* appear generally to go both ways, as handily backward as forward. A particle and its antiparticle can pop out of the vacuum, and then annihilate, leaving behind only radiation. If you ran this movie backward, you couldn't tell the difference. As Wheeler explains it, subatomic particles such as electrons have no past, no history. You cannot tell anything about where an electron has been. Past and future are the same. "The electron pays for its freedom to move forward and backward in time by remembering neither future nor past. We remember the past and are trapped in one-way motion through time."

Physicists can't quite explain this difference except to say that one-way time is a statistical property, like the fact that tossed coins will tend to land heads up 50 percent of the time given enough throws. And statistical properties only emerge on large scales.

Whatever the reason, time does not appear to be symmetrical, while space is. This flies in the face of Einstein's special relativity, which puts time and space on equal footing. Time and space are supposed to be interchangeable, like two sides of a coin. Woven together into the larger fabric of spacetime, the two form a single entity. This certainly suggests that paradoxes in time should run parallel to paradoxes of space. And yet, they seem to be quite distinct.

Space, for example, can move in many different directions and dimensions; space can come unglued. But if time becomes unglued, cause and effect fall to pieces. (See Chapter 6.) A cause has to happen *before* an effect—by definition. But if time gets tangled, can cause and effect change places? And if they do, does the idea of causation make any sense?

For now, no one knows—except that whatever notion of causality emerges from a new understanding of time will be quite different from what it is today. "It's more subtle," says Greene, who then admits, "That's a fancy way of saying I don't understand what it is."

There's hope, however, on both the experimental and theoretical fronts. For perhaps the first time, physicists are currently probing time's mysteries head-on.

For example, a series of experiments, ongoing and planned, will probe certain cracks in the perfect symmetry of subatomic time that were first discovered in the 1960s. Previously, it had been thought that all subatomic processes were perfectly symmetrical with respect to time: forward and backward were one and the same.

In the 1960s, however, physicists discovered that in certain rare cases, forward and backward do not necessarily look the same. These violations of normal submicroscopic goings-on are part of a larger picture in which particles and antiparticles also change places. Indeed, these very violations might shed light on why the universe is dominated by matter—because time reversal and matter/antimatter reversal are connected.

As for theorists, they will continue to dial back the clock of the universe to times when space and time alike appear to break down—the realm of quantum foam and, possibly, strings. It is, of course, impossible to return to time zero—even in thought experiments. To go this far takes physicists far beyond anything they understand, so there would be no reason to trust such extrapolations. As Guth points out, "Our knowledge is very shaky here."

Still, one can imagine a situation where the same kinds of conditions would prevail that must have existed at the onset of time—a time and place where everything in the universe is condensed to a single infinitely small point, as in the inside of a black hole. The very fact that space and time must break down in such realms probably makes them good places to look for clues.

It would be fitting, indeed, if one of the primary properties that distinguishes something from nothing—time itself—murmurs its secrets from the zero hour of the universe.

SOMETHING WONDERFUL

Nothing is too wonderful to be true.
—Michael Faraday

So why is it that nature is almost perfectly symmetrical—but not quite? Why do so many physical quantities approach zero so closely, only to never quite arrive? Why did matter nudge ever so slightly ahead of antimatter in the chaos of mutual annihilation that reigned in the early years of the universe? Why are neutrinos so tiny? Why does time run one way?

What, in other words, is the reason for the pervasive imperfection that makes everything possible?

Richard Feynman raises this question in his *Lectures on Physics.* "No one has any idea why," he says.

He suggests, however, that the answer might have something to do with the fact that perfect symmetry is simply too beautiful to be true. He then tells the story of a gate in Neiko, Japan, known as the most beautiful gate in the country. The gate is elaborately carved with designs that are perfectly symmetrical except for one small element which is carved upside down. It was carved upside down deliberately, so the story goes, in order to mar the perfection—so that the gods would not become jealous and angry.

"We might like to turn the idea around," Feynman concludes, "and think the true explanation of the near symmetry of nature is this: that God made the laws only nearly symmetrical so that we should not be jealous of His perfection."

Luckily for all of us, nothing may be too beautiful to be true, but everything else is not.

Supporting Cast

Sidney Coleman is a physicist at Harvard University. Known among his colleagues as the man who knows more about nothing than anyone else, he has written several seminal papers on the properties of the vacuum.

Savas Dimopoulos is a physicist at Stanford University whose recent work on "large" extra dimensions has opened the doors to the possibility that bizarre new physics might be closer—or at least more detectable—than previously thought.

Martin Gardner is so admired and beloved for his (literally) magical skills in writing about mathematics and exposing dubious science that every two years, in January, hundreds of his fans gather from all over the world to honor him. My personal favorite of his many published works is *The Annotated Alice: The Definitive Edition.*

Brian Greene, a mathematician and physicist at Columbia University, is the author of the best-selling book on string theory, *The Elegant Universe: Superstrings, Hidden Dimensions, and the Quest for the Ultimate Theory.* He plays himself in the movie *Frequency.*

David Gross is director of the Institute for Theoretical Physics at the University of California at Santa Barbara. Along with Frank Wilczek, he discovered the strange nature of the force between quarks that keeps them permanently trapped.

Alan Guth of the Massachusetts Institute of Technology is known for one of the single most influential theories in cosmology—the extraordinary inflation of the universe in its earliest moments. He explains how to cook up universes from nothing in his book *The Inflationary Universe: The Quest for a New Theory of Cosmic Origins.*

Stephen Hawking's many contributions to physics include the startling theoretical discovery that black holes "ain't black," as he puts it, but can radiate energy, and eventually even evaporate. He is Lucasian Professor of Mathematics at the University of Cambridge, and the author of *A Brief History of Time.*

Gary Horowitz is a physicist at the University of California at Santa Barbara. His work focuses on black holes and string theory.

Thomas Humphrey worked as a physicist at Fermilab before coming to San Francisco to help physicist Frank Oppenheimer found the Exploratorium in the early 1970s. Sculptor, musician, and master teacher, he is currently senior scientist at the Exploratorium.

Rocky (Edward W.) Kolb is a cosmologist at the University of Chicago and founder of the theory group at Fermi National Accelerator Laboratory in Batavia, Illinois. Known as one of Fermilab's foremost humorists, he is the author of *Blind Watchers of Sky: The People and Ideas That Shaped Our Idea of the Universe.*

Leon Lederman won the Nobel Prize in 1988 for discovering the muon neutrino, a species of particle often referred to as a spinning nothing. He is director emeritus of Fermilab as well as the author, with Dick Teresi, of probably the funniest book about particle physics: *The God Particle: If the Universe is the Answer, What is the Question?*

Helen Quinn is a physicist at the Stanford Linear Accelerator Laboratory. In 1998, she was elected to the American Academy of Arts and Sciences for her "distinguished contributions to the profession."

Johann Rafelski is a physicist at the University of Arizona, Tucson, who has devoted much of his career to studying properties of the vacuum. He is the author, along with Berndt Muller, of *The Structured Vacuum: Thinking about Nothing.*

Lisa Randall is a physicist at Massachusetts Institute of Technology and Princeton whose work on large extra dimensions has been the center of attention at many recent meetings on string theory. Her work suggests that unseen dimensions could be infinite in extent, rather than tiny and curled up.

Nathan Seiberg is a physicist at the Institute for Advanced Study, in Princeton, New Jersey. Along with Edward Witten, he developed the Seiberg-Witten equations which described new ways of thinking about four-dimensional space and helped propel string theory to the center stage of physics.

Andrew Strominger is a string theorist at Harvard University; he is known for showing how strings and black holes can appear to be different aspects of the same thing, like water changing to ice.

Leonard Susskind is a physicist at Stanford University. He is one of the founders of string theory and more recently one of the developers of the holographic universe—the notion that all the information in a higher-dimensional reality can be encoded on a lower-dimensional surface.

Michael Turner is a cosmologist at the University of Chicago and Fermi National Accelerator Laboratory. His presentations are famous for their original artwork. His art has appeared in the *New York Times,* and he has had a one-man show at the gallery of the Center for Particle Astrophysics in Berkeley.

Frank Wilczek is a physicist at the Institute for Advanced Study, in Princeton, New Jersey, and author, along with Betsy Devine, of *Longing for the Harmonies: Longing for the Harmonies: Themes and Variations from Modern Physics.* He writes frequently for *Physics Today* magazine, often about the vacuum.

Edward Witten is known by his colleagues as "the pope of string theory." A physicist at the Institute for Advanced Study, in Princeton, New Jersey, he is currently spending a year at the California Institute of Technology, in Pasadena.

Bibliography

Ascher, Marcia. *Ethnomathematics: A Multicultural View of Mathematical Ideas.* Pacific Grove, Calif.: Brooks/Cole Pub., 1991.

Aveni, Anthony. *Conversing with the Planets: How Science and Myth Invented the Cosmos.* New York: Kodansha International, 1994.

Barker, David. "Untitled: Art, Music, and Nothingness." *Exploratorium Magazine: Exploring Nothing,* vol. 13, no. 4 (winter 1989).

Barrow, John D. *Impossibility: The Limits of Science and the Science of Limits.* Oxford: Oxford University Press, 1998.

Bell, Eric Temple. *The Magic of Numbers.* 1946. Reprint, New York: Dover Publications, Inc., 1991.

Berlinski, David. *A Tour of the Calculus.* New York: Vintage Books, 1995.

Boole, George. *Investigation of the Laws of Thought.* 1958. Reprint, New York: Dover Publications, Inc., 1973.

Boorstin, Daniel. *The Discoverers.* New York: Random House, 1983.

Brown, Kurt, ed. *Verse and Universe: Poems about Science and Mathematics.* Minneapolis, Minn.: Milkweed Editions, 1998.

Bunch, Bryan H., et al. *Mathematical Fallacies and Paradoxes.* 1982. Reprint, New York: Dover Publications, Inc., 1997.

Caldwell, David O., ed. *COSMO-98: Second International Workshop on Particle Physics and the Early Universe.* Woodbury, N.Y.: American Institute of Physics, 1998.

Capra, Fritjof. *The Tao of Physics: An Exploration of the Parallels between Modern Physics and Eastern Mysticism.* Boston: Shambhala Publications, 1975.

Chagme, Karma. *A Spacious Path to Freedom: Practical Instructions on the*

Union of Mahamudra and Atiyoga. Translated by B. Alan Wallace. Ithaca, N.Y.: Snow Lion Publications, 1998.

Cole, K. C. "Prof. Edward Tryon Talks (at Length) about Nothing." *Hunter Magazine,* vol. 4 (July 1985).

Coleman, Sidney. "Why There Is Nothing Rather Than Something: A Theory of the Cosmological Constant." *Nuclear Physics B.,* vol. 310, nos. 3–4 (December 12, 1988): 643.

Conway, John H., and Richard K. Guy. *The Book of Numbers.* New York: Copernicus, 1996.

Cornish, Neil J., and Jeffrey R. Weeks. "Measuring the Shape of Space." *Notices of the American Mathematical Society,* vol. 45, no. 11 (December 1998).

Crease, Robert P., and Charles C. Mann. *The Second Creation: Makers of the Revolution in Twentieth-Century Physics.* Reprint, New Brunswick, N.J.: Rutgers University Press, 1996.

Cronin, Anthony. *Samuel Beckett: The Last Modernist.* New York: HarperCollins, 1997.

Dantzig, Tobias. *Number: The Language of Science.* 4th ed. New York: Free Press, 1967.

Davies, Paul. *The Last Three Minutes: Conjectures about the Ultimate Fate of the Universe.* New York: Basic Books, 1994.

——, and John Gribbin. *The Matter Myth: Dramatic Discoveries that Challenge Our Understanding of Physical Reality.* New York: Simon & Schuster, 1992.

DeLuccia, Frank, and Sidney Coleman. "Gravitational Effects on and of Vacuum Decay." *Physical Review D,* vol. 21, no. 12 (1980): 3305.

Devlin, Keith. The *Language of Mathematics: Making the Invisible Visible.* New York: W. H. Freeman & Company, 1998.

Dewdney, A. K. *A Mathematical Mystery Tour: Discovering the Truth and Beauty of the Cosmos.* New York: John Wiley & Sons, Inc., 1999.

Dunham, William. *The Mathematical Universe: An Alphabetical Journey through the Great Proofs, Problems, and Personalities.* New York: John Wiley & Sons, Inc., 1994.

Eddington, Sir Arthur. *Space, Time, and Gravitation: An Outline of the General Relativity Theory.* New York: Harper & Row, 1959.

Einstein, Albert. *Essays in Physics.* New York: Philosophical Library, 1950.

——. *Ideas and Opinions.* Edited by Carl Seelig. Translated by Sonja Bargmann. New York: Crown Publishers, Inc., 1982.

——, and Leopold Infeld. *The Evolution of Physics*: from Early Concepts to Relativity and Quanta. New York: Simon & Schuster, 1966.

Epstein, Mark. *Going to Pieces without Falling Apart: A Buddhist Perspective on Wholeness.* New York: Broadway Books, 1998.

Exploratorium Magazine. Exploring: Nothing, vol. 13, no. 4 (winter 1989).

Ferguson, Kitty. *Measuring the Universe: Our Historic Quest to Chart the Horizons of Space and Time.* New York: Walker & Company, 1999.

Ferris, Timothy. *The Whole Shebang: A State of the Universe(s) Report.* New York: Simon & Schuster, 1997.

Feynman, Richard. *The Character of Physical Law.* Cambridge, Mass.: M.I.T. Press, 1965.

——, et al. *The Feynman Lectures on Physics.* Reading, Mass.: Addison-Wesley, 1963–1965.

Flegg, Graham. *Numbers: Their History and Meaning.* New York: Schocken Books, 1983.

Friedberg, Richard. *An Adventurer's Guide to Number Theory.* 1968. Reprint, New York: Dover Publications, Inc., 1994.

Gamow, George. *Mr. Tompkins in Paperback.* Cambridge: Cambridge University Press, 1965.

Gardner, Martin. "More Ado about Nothing." In *Mathematical Magic Show.* Washington, D.C.: The Mathematical Association of America, 1990.

——. "Zero-Point Energy and Harold Puthoff." *Skeptical Inquirer,* vol. 22, no. 3 (May/June 1998): 13.

Garland, Trudi Hammel. *Fascinating Fibonaccis: Mystery and Magic in Numbers.* Palo Alto, Calif.: Dale Seymour Publications, 1987.

Genz, Henning. *Nothingness: The Science of Empty Space.* Translated by Karin Heusch. Reading, Mass.: Perseus Books, Helix Books, 1999.

Glanz, James, "Breakthrough of the Year: Cosmic Motion Revealed." *Science,* vol. 282, no. 5397 (December 18, 1998): 2156.

Gleiser, Marcelo. *The Dancing Universe: From Creation Myths to the Big Bang.* New York: Dutton Books, 1997.

Gould, Stephen Jay. *Questioning the Millennium: A Rationalist's Guide to a Precisely Arbitrary Countdown.* New York: Harmony Books, 1997.

——. *Time's Arrow, Time's Cycle: Myth and Metaphor in the Discovery of Geological Time.* Cambridge, Mass.: Harvard University Press, 1987.

Greene, Brian. *The Elegant Universe: Superstrings, Hidden Dimensions, and the Quest for the Ultimate Theory.* New York: W. W. Norton & Company, 1999.

Gregory, Richard. *The Intelligent Eye.* New York: McGraw-Hill, 1970.

——. "Science through Play." In *Science Today: Problem or Crisis?* Edited by Ralph Levinson and Jeff Thomas. London: Routledge, 1997.

——, ed. *The Oxford Companion to the Mind.* Oxford: Oxford University Press, 1987.

Gross, David J. "The Role of Symmetry in Fundamental Physics" *Proceedings of the National Academy of Sciences of the United States,* vol. 93, no. 25 (December 10, 1996):14256.

Gullberg, Jan. *Mathematics: From the Birth of Numbers.* New York: W. W. Norton & Company, 1997.

Guth, Alan H. "Inflationary Models and Connections to Particle Physics." In *The Pritzker Symposium on the Status of Inflationary Cosmology.* Edited by N. Pritzker and M. S. Turner. Forthcoming, Chicago: University of Chicago Press, 2001. Los Alamos National Laboratory Archives: http://xxx.lanl.gov/abs/astro-ph/002188.

Guth, Alan H. *The Inflationary Universe: The Quest for a New Theory of Cosmic Origins.* Reading, Mass.: Addison-Wesley, Helix Books, 1997.

Hawking, Stephen. *Black Holes and Baby Universes and Other Essays.* New York: Bantam Books, 1980.

——. *A Brief History of Time: From the Big Bang to Black Holes.* New York: Bantam Books, 1988.

——, and Roger Penrose. *The Nature of Space and Time.* Princeton, N.J.: Princeton University Press, 1996.

Hersh, Reuben. *What Is Mathematics, Really?* New York: Oxford University Press, 1997.

Hewitt, Paul G. *Conceptual Physics for Parents and Teachers: Mechanics.* Newburyport, Mass.: Focus Information Group, 1998.

Hoffmann, Banesh. *Relativity and Its Roots.* 1983. Reprint, Mineola, N.Y.: Dover Publications, Inc., 1999.

Hogben, Lancelot. *Mathematics for the Million: How to Master the Magic of Numbers.* 4th ed. New York: W. W. Norton & Company, 1968.

Holton, Gerald. *Einstein, History, and Other Passions: The Rebellion Against Science at the End of the Twentieth Century.* Reading Mass.: Addison-Wesley Publishing, 1996.

———. "I. I. Rabi as Educator and Science Warrior." *Physics Today,* vol. 52, no. 9 (September 1999): 37.

———, and Yehuda Elkana, eds. *Albert Einstein: Historical and Cultural Perspectives.* 1982. Reprint, Mineola, N.Y.: Dover Publications, Inc., 1997.

Huggett, Nick, ed. *Space from Zeno to Einstein: Classic Readings with a Contemporary Commentary.* Cambridge, Mass.: MIT Press, 1999.

Jacobson, Lyle D. *Discovery of Nothing: A New Interpretation of Reality.* Dubuque, Iowa: Kendall/Hunt Publishing Company, 1998.

Jammer, Max. *Concepts of Space: The History of Theories of Space in Physics.* 1945. Reprint, New York: Dover Publications, Inc., 1994.

Kaku, Michio, and Jennifer Trainer Thompson. *Beyond Einstein: The Cosmic Quest for the Theory of the Universe.* New York: Anchor, 1987.

Kaplan, Robert. *The Nothing That Is: A Natural History of Zero.* New York: Oxford University Press, 1999.

King, Jerry P. *The Art of Mathematics.* New York: Ballantine Books, 1992.

Knowlson, James. *Damned to Fame: The Life of Samuel Beckett.* London: Bloomsbury Publishing, 1996.

Krauss, Lawrence M. *Beyond "Star Trek": Physics from Alien Invasions to the End of Time.* New York: Basic Books, 1997.

———. *Fear of Physics: A Guide for the Perplexed.* New York: BasicBooks, 1993.

———. *Quintessence: The Mystery of the Missing Mass in the Universe.* Revised edition of *The Fifth Essence: The Search for Dark Matter in the Universe.* New York: Basic Books, 2000.

Krieger, Martin H. *Doing Physics: How Physicists Take Hold of the World.* Bloomington: Indiana University Press, 1992.

Lange, Andrew E. "Understanding Creation: Cosmology at the Dawn of the Twenty-first Century."

Lederman, Leon, with Dick Teresi. *The God Particle: If the Universe Is the Answer, What Is the Question?* Boston: Houghton Mifflin, 1993.

Lethem, Jonathan. *As She Climbed across the Table.* New York: Vintage Books, 1997.

Lightman, Alan P. *Dance for Two: Selected Essays:* New York, Pantheon Books, 1996.

Lineweaver, Charles H. *What Is the Universe Made Of? How Old Is It?* Los Alamos National Laboratory Archives: http://xxx.lanl.gov/abs/astro-ph/9911294

Lucretius, *On the Nature of the Universe.* Translated by R. E. Latham. London: Penguin Books, 1994.

Maddox, John. *What Remains to be Discovered: Mapping the Secrets of the Universe, the Origins of Life, and the Future of the Human Race.* New York: Martin Kessler Books, 1998.

March, Robert H. *Physics for Poets.* 4th ed. New York: McGraw-Hill, Inc., 1996.

Menninger, Karl. *Number Words and Number Symbols: A Cultural History of Numbers.* Translated by Paul Broneer. New York: Dover Publications, Inc., 1992.

Morrison, Philip, and Phylis Morrison. *Powers of Ten.* New York: Scientific American Library, 1982.

——. *The Ring of Truth.* New York: Random House, 1987.

Motz, Lloyd, and Jefferson Hane Weaver. *The Story of Mathematics.* New York: Plenum Press, 1993.

Muir, Jane. *Of Men and Numbers: The Story of the Great Mathematicians.* New York: Dover Publications, Inc., 1996.

Nerlich, Graham. *The Shape of Space.* 2d ed. Cambridge: Cambridge University Press, 1994.

Newton, Isaac. "The New Theory about Light and Color." In *Newton's Philosophy of Nature: Selections from His Writings.* Edited by H. S. Thayer. New York: Hafner Publishing Company, 1953.

——. "Hypothesis Touching on the Theory of Light and Colors." In *Newton's Philosophy of Nature: Selections from His Writings.* Edited by H. S. Thayer. New York: Hafner Publishing Company, 1953.

Newton, Roger G. *The Truth of Science: Physical Theories and Reality.* Cambridge, Mass.: Harvard University Press, 1997.

Nishitani, Keiji. *Religion and Nothingness.* Translated by Jan Van Bragt. Berkeley: University of California Press, 1982.

Norretranders, Tor. *The User Illusion: Cutting Consciousness Down to Size.* Translated by Jonathan Sydenham. New York: Viking, 1998.

Overbye, Dennis. *Lonely Hearts of the Cosmos: The Scientific Quest for the Secret of the Universe.* New York, HarperCollins Publishers, 1991.

Paulos, John Allen. *A Mathematician Reads the Newspaper.* New York: BasicBooks, 1995.

Pickover, Clifford. *The Keys to Infinity.* New York: John Wiley & Sons, Inc., 1995.

Plant, Sadie. *Zeroes + Ones: Digital Women + The New Technoculture.* New York: Doubleday, 1997.

Poundstone, William. *Labyrinths of Reason: Paradox, Puzzles, and the Frailty of Knowledge.* New York: Doubleday, 1988.

Rafelski, Johann, and Berndt Muller. *The Structured Vacuum: Thinking about Nothing.* Germany: Verlag Harri Deutsch, 1985.

Ramachandran, V. S., and Sandra Blakeslee. *Phantoms in the Brain.* New York: William Morrow & Company, Inc., 1998.

Rees, Martin. *Before the Beginning: Our Universe and Others.* Reading, Mass.: Addison-Wesley, 1997.

Reid, Constance. *From Zero to Infinity: What Makes Numbers Interesting.* 4th ed. Washington, D.C.: The Mathematical Association of America, 1992.

Ridley, B. K. *Time, Space, and Things.* 3d ed. Cambridge: Cambridge University Press, 1995.

Rotman, Brian. *Signifying Nothing: The Semiotics of Zero.* Stanford, Calif.: Stanford University Press, 1987.

Sacks, Oliver. *A Leg to Stand On.* 1st Touchstone ed. New York: Simon & Schuster, Inc., 1998.

——. "Nothingness." In *The Oxford Companion to the Mind.* Edited by Richard L. Gregory. Oxford: Oxford University Press, 1987.

Sartre, Jean-Paul. *Being and Nothingness: An Essay on Phenomenological Ontology.* Translated by Hazel E. Barnes. Reprint, New York: Washington Square Press, 1993.

Schimmel, Annemarie. *The Mystery of Numbers.* Oxford: Oxford University Press, 1993.

Siegfried, Tom. *The Bit and the Pendulum: From Quantum Computing to M Theory—the New Physics of Information.* New York: John Wiley & Sons, Inc., 2000.

Smolin, Lee. *The Life of the Cosmos.* New York: Oxford University Press, 1997.

——. "Toward a Background Independent Approach to M Theory." Published by the Center for Gravitational Physics and Geometry, The Pennsylvania State University, University Park, PA 16802-6300.

Stachel, John, ed. *Einstein's Miraculous Year: Five Papers That Changed the Face of Physics.* Princeton, N.J.: Princeton University Press, 1998.

Stewart, Ian. *From Here to Infinity: A Guide to Today's Mathematics.* Revised edition. New York: Oxford University Press, 1996.

——. *Life's Other Secret: The New Mathematics of the Living World.* New York: John Wiley & Sons, Inc., 1998.

——. *The Magical Maze: Seeing the World through Mathematical Eyes.* New York: John Wiley & Sons, Inc., 1998.

——. *Nature's Numbers: The Unreal Reality of Mathematical Imagination.* New York: BasicBooks, 1995.

——. "Zero, Zilch, and Zip." *New Scientist,* vol. 158, no. 4 (April 1998): 40.

Teresi, Dick. "Zero." *The Atlantic Monthly,* July 1997: 88.

Thayer, H. S., ed. *Newton's Philosophy of Nature: Selections From His Writings.* New York: Hafner Publishing Company, 1953.

Thorne, Kip S. *Black Holes and Time Warps: Einstein's Outrageous Legacy.* New York: W. W. Norton & Company, 1994.

Trefil, James S. " 'Nothing' May Turn Out to Be the Key to the Universe: Cosmological Speculations on the Nature of Perfect Vacuum and Elementary Particle Interactions." *Smithsonian Magazine,* vol. 12, no. 9 (1981): 142.

——. *101 Things You Don't Know About Science and No One Else Does Either.* Boston: Houghton Mifflin Company, 1996.

Turner, Michael. "The Cosmology of Nothing." In *Vacuum and Vacua: The Physics of Nothing.* Proceedings of the International Society of Subnuclear Physics. The Subnuclear Series, vol. 33. Edited by A. Zichichi. Geneva, Switzerland: European Physics Society, 1996.

Veneziano, Gabriele. "Challenging the Big Bang." *CERN Courier,* March 1999.

Vogt, Erich. "The Art and Science of Finding Nothing." From *Rare Decay Symposium.* Edited by D. Bryman. Singapore: World Scientific Publishers, 1988.

Von Baeyer, Hans Christian. *The Fermi Solution: Essays on Science.* New York: Random House, Inc., 1993.

Wallace, B. Alan. *Choosing Reality: A Buddhist View of Physics and the Mind.* Ithaca, N.Y.: Snow Lion Publications, 1996.

Weinberg, Steven. *Dreams of a Final Theory: The Scientist's Search for the Ultimate Laws of Nature.* New York: Pantheon Books, 1992.

——. *The First Three Minutes: A Modern View of the Origin of the Universe.* Updated edition. New York: Basic Books, 1993.

Wells, David. *The Penguin Dictionary of Curious and Interesting Numbers.* New York: Penguin Books, 1986.

Wertheim, Margaret. *The Pearly Gates of Cyberspace: A History of Space from Dante to the Internet.* New York: W. W. Norton & Company, 1999.

Wheeler, John A. *At Home in the Universe.* New York: Springer-Verlag, 1996.

——, and Kenneth William Ford. *Geons, Black Holes, and Quantum Foam: A Life in Physics.* New York: W. W. Norton & Company, 1998.

Wilczek, Frank. "The Cosmic Asymmetry between Matter and Anti-Matter. *Scientific American,* vol. 243, no. 6 (December 1980): 82.

——. "The Persistence of Ether." *Physics Today,* vol. 52, no. 1 (January 1999): 11.

——. "Reference Frame; Mass without Mass I: Most of Matter." *Physics Today,* vol. 52, no. 11 (November, 1999): 11.

Wilson, Fred. *Viewing the Invisible.* Melbourne, NSW, Australia: Ian Potter Museum of Art, 1998.

Witten, Edward. "Duality, Spacetime and Quantum Mechanics." *Physics Today,* May 1997: 28.

——. "Reflections on the Fate of Spacetime." *Physics Today.* April 1996: 24.

Yam, Philip. *Exploiting Zero-Point Energy. Scientific American,* vol. 277, no. 6 (December 1997): 82.

Zajonc, Arthur. *Catching the Light: The Entwined History of Light and Mind.* New York: Oxford University Press, 1995.

Zichichi, Antonino, ed. *Vacuum and Vacua: The Physics of Nothing.* Proceedings of the International School of Subnuclear Physics. Singapore: World Scientific Publishing Co., 1996.

Index